Caspian Energy Resources

Implications for the Arab Gulf

Caspian Energy Resources

Implications for the Arab Gulf

THE EMIRATES CENTER FOR STRATEGIC STUDIES AND RESEARCH

CASPIAN ENERGY RESOURCES
Implications for the Arab Gulf

Published by
The Emirates Center for Strategic Studies and Research
PO Box 4567
Abu Dhabi
United Arab Emirates

e-mail: root@ecssr.ac.ae
http: //www.ecssr.ac.ae

Distributed by: I. B. Tauris & Co Ltd
6 Salem Road, London W2 4B
175 Fifth Avenue, New York NY10010

ISBN 0 86372 278 4

First Edition

British Library Cataloguing-in-Publication Data
A catalogue record for this book is available from the British Library

Printed in Lebanon

The Emirates Center for Strategic Studies and Research

The Emirates Center for Strategic Studies and Research (ECSSR) is an independent research institution dedicated to the promotion of professional studies and educational excellence in the United Arab Emirates (UAE), the Gulf and the Arab world. Since its establishment in Abu Dhabi in 1994, ECSSR has served as a focal point for scholarship on political, economic and social matters. Indeed, ECSSR is at the forefront of analysis and commentary on Arab affairs.

The Center provides a forum for the scholarly exchange of ideas by hosting conferences and symposia, organizing workshops, sponsoring a lecture series and publishing original and translated books and research papers. ECSSR also has an active fellowship and grant program for the writing of scholarly books and for the translation, into Arabic, of works relevant to the Center's mission. Moreover, ECSSR has a large library including rare and specialized holdings, and a state-of-the-art technology center, which has developed an award-winning website that is a unique and comprehensive source of information on the Gulf.

Through these and other activities, ECSSR aspires to engage in mutually beneficial professional endeavors with comparable institutions worldwide, and to contribute to the general educational and scientific development of the UAE.

Contents

Tables and Figures

TABLES

FIGURES

Abbreviations

ACG	Azeri-Chirag-Güneshli
AIOC	Azerbaijan International Operating Company
API	American Petroleum Institute
b/d	barrels per day
BBC	British Broadcasting Corporation
bcm	billion cubic meters
bn/b	billion barrels
boe	barrels of oil equivalent
BOTAŞ	Turkish State Pipeline Corporation
BP	British Petroleum
CASCO	Caspian Sea Cooperation Organization
CERA	Cambridge Energy Research Associates
CGES	Centre for Global Energy Studies
CIPCO	Caspian International Petroleum Company
CIS	Commonwealth of Independent States
CNN	Cable News Network
CPC	Caspian Pipeline Consortium
CPSU	Communist Party of the Soviet Union
CSIS	Center for Strategic and International Studies
DOE	Department of Energy (US)
EBRD	European Bank for Reconstruction and Development
ECO	Economic Cooperation Organization
ECT	Energy Charter Treaty
EIA	Energy Information Administration
ESAI	Energy Security Analysis, Inc.
EXIMBANK	Export–Import Bank (US)
FDI	foreign direct investment
FIOC	First International Oil Company
FSU	former Soviet Union
GATT	General Agreement on Tariffs and Trade
GDP	gross domestic product
GCC	Gulf Co-operation Council
GNP	gross national product

IDE	Institute of Developing Economies
IEA	International Energy Agency
IMF	International Monetary Fund
IPA	Islamic Party of Azerbaijan
IPC	Iraq Petroleum Company
KGB	Komitet Gossudarrstvennoi Bezopastnosti (State Security Police)
mbpd	million barrels per day
MIT	Massachusetts Institute of Technology
mt	million tons
MTCR	Missile Technology Control Regime
NAOC	North Absheron Operating Company
NATO	North Atlantic Treaty Organization
NIS	newly independent states
OAPEC	Organization of Arab Petroleum Exporting Countries
ODIHR	Office for Democratic Institutions and Human Rights
OECD	Organisation for Economic Co-operation and Development
OKIOC	Offshore Kazakhstan International Operating Company
OMI	*Oil Market Intelligence*
OOM	Oman Oil Company
OPEC	Organization of Petroleum Exporting Countries
OPIC	Overseas Private Investment Corporation
OSCE	Organization for Security and Co-operation in Europe
PIW	*Petroleum Intelligence Weekly*
PKK	Kurdistan Workers' Party
PRT	Petroleum Revenue Tax
PSC	Production Sharing Contract
SAIS	School for Advanced International Studies (US)
SOCAR	State Oil Company of the Azerbaijan Republic
sq. km.	square kilometers
SSR	Soviet Socialist Republic
tcf	trillion cubic feet
TDA	Trade and Development Agency (US)
UAE	United Arab Emirates
UK	United Kingdom
US	United States (of America)
USSR	Union of Soviet Socialist Republics
WTO	World Trade Organization

Introduction

Robert Ebel

History does have a way of repeating itself. Observers of the oil industry long ago learned, to their advantage, that those forces which shape decision-making today were just as influential in years past. Those factors which encouraged or discouraged early investors clearly have not become relics of their time. Although the oil industry may have grown in size and in sophistication, in quality of performance, and in importance to the political and economic well-being of nations, it is still the access to oil which continues to drive producers and consumers alike.

Since the very beginning, the question of how to move crude oil from its place of production to refineries, and then the finished products to the consumer, has confronted the industry. In the early days in the United States, crude oil was stored in wooden barrels, piled up in horse-drawn wagons and taken to the nearest railroad for transportation. The cost of moving the crude oil by wagon to the nearest rail stop usually exceeded the cost of the much longer rail shipment.

Then, to break the monopoly held by teamsters who drove these wagons, and to cut costs as well, wooden pipelines were laid to connect the producing wells with nearby railroads.

During this time, John D. Rockefeller was strengthening his position in the American oil industry, initially through the acquisition and expansion of refining capacity. Eventually, by the early 1870s, Rockefeller had control over the largest refinery group in the world. It was a time when the supply of crude oil was growing much faster than demand, causing oil prices to fall sharply. Rockefeller used this opportunity to acquire assets at very low prices and came to dominate not only refining capacity but transportation as well.

One last effort was made by a group of independent producers to break Rockefeller's hold. They undertook construction of the world's first long-distance pipeline, 110 miles in length. Four others followed shortly. The success of these pipelines changed the oil industry forever. No longer was crude oil limited to nearby markets. Now, markets distant from sources of production could benefit from this new form of energy as well. The challenge to Rockefeller's Standard Oil was successful and it was made possible by independent access to markets.

Rockefeller quickly understood the implications of long-distance transport of crude oil and petroleum products by pipeline. Not only was it far cheaper than rail, but if he could add pipelines to his control of production and refining, the oil industry would be completely within his grasp. For, just as the independents had realized, pipelines were the key to a profitable future. The lesson learned was simple: crude oil will not be produced if it cannot be moved to market.

This lesson has not been lost on today's oil industry. Azerbaijan and Central Asia stand today where these independent producers and Rockefeller stood a century ago. In the absence of pipeline capacity to move new volumes of crude oil and natural gas to Western markets, production cannot be expanded and the prospect of a prosperous future would be postponed indefinitely.

Prior to the Communist revolution in 1917, Western oil interests and investments in Russia were considerable, and were dominated by the three Nobel brothers, the Rothschilds, Henri Deterding of Royal Dutch and Marcus Samuel of Shell. By 1897 the Russian oil fields – just twelve square miles around Baku – delivered more than 45 percent of the world's crude oil. And a few years later, by the end of the century, Russian wells were providing more than half the total world production. But the Russian leadership lost ground rather quickly, as the United States took the lead, building on oil discoveries in Texas, Louisiana and Oklahoma.

Despite the absence of a broad industrial base, the Russian economy absorbed much of the domestic oil supply. The amounts available for export were further constrained by the transportation and freight policy of the Russian government, which had imposed punitive tariffs on those volumes of oil moving from production fields around Baku to the export outlet of Batumi on the Black Sea. These tariffs in effect deliberately diverted the flow of oil inland, to the extent that in the peak year of oil exports from pre-revolutionary Russia, only some 17 percent was sold abroad.

These earlier tactics warrant attention, for the need for contributions to the budget, the rationale behind these tariffs, pales in comparison with the budgetary requirements of Russia today. In the very early days, Russian oil was moved to refineries just as it was in the United States, in wooden barrels piled in carts pulled by horses or mules, a method that was too costly, slow, and not very dependable. Russia eagerly sought information pertaining to the steel pipelines built in the state of Pennsylvania. What diameters were the pipelines, how were the individual pipe sections joined together and how was leakage kept to a minimum? The first steel pipeline was laid beginning in 1877, from the Balakhany oil field to the refinery in Black Town, where the oil workers lived. The going was not easy for Baku officials were reluctant to give the necessary transit permission, displaced workers attacked the pipeline and armed guards had to be hired.

The Nobel brothers then followed with an expanding fleet of river barges, tankers and rail tank cars to carry crude oil to refineries and petroleum products to customers. Rail tank cars moved along the Baku–Batumi route which had been taken from Turkey in 1878. Competitors, financed by the Rothschilds, built a second Baku–Batumi rail line. Batumi was turned into a major port for the export of petroleum products on an international scale.

What logically would follow? The Baku–Batumi rail line was limited in its carrying capacity and tracks were often washed away by heavy rains. Nobel had built a short pipeline along the rail route in 1889, to ease the oil flow through, not over, a mountain at Suram. Seventeen years later, in 1906, the Russian government completed a 560-mile, eight-inch pipeline between Baku and Batumi, at that time the longest in the world.

Soviet literature rarely discussed the pre-revolutionary history of the Russian oil industry and the many accomplishments achieved during those years. When it did, that history was dispensed with rather quickly, with emphasis placed on worker exploitation and the difficult living and working conditions forced upon them. The decline in oil output was significant, however, as production fell from 225,000 barrels per day (b/d) in 1901 to 176,000 b/d in 1917. In reality, this was not all that unexpected, given the impact of riots, civil war and World War I.

It is the power of oil that brings nations, multinational oil companies and entrepreneurs to the Caspian Basin. Just what is that power? Oil is high-profile stuff, for it fuels much more than automobiles and airplanes. Oil fuels military power, national treasuries and international politics. Because of this, it is no longer merely a commodity to be bought and sold within the confines of traditional energy supply and demand balances. Rather, it has been transformed into a determinant of well-being, of national security and international power for those who possess this vital resource, and the converse for those who do not.

For the United States, the support of Caspian oil development began as an outgrowth of a national energy policy that calls for the expansion of oil production outside the United States but away from the Arab Gulf. This policy implicitly recognizes that the United States is dependent on cheap foreign oil and that its reliance on oil imports will only increase over the coming years. Any importing nation like the United States seeks to enhance its national security through diversity among sources of supply, and Caspian oil clearly met that requirement. The US policy subsequently evolved over time to one which came to embrace the following, not necessarily in any order of priority:

- enhancing commercial opportunities for the United States and American companies
- strengthening the political and financial independence of Azerbaijan, Kazakhstan and Turkmenistan, among others
- building economic linkages (e.g. pipelines) between these newly independent states which hopefully would mitigate regional conflicts.

During the zenith of the Soviet crude oil and natural gas industries, that is, the 1970s and most of the 1980s, Central Asia and Azerbaijan played vastly different roles. Central Asia, largely Turkmenistan, was a major contributor to supplies of natural gas, with its share exceeding 24 percent as early as 1970 and rising to just under 33 percent by 1975, a scant five years later. Conversely, Azerbaijan's share of Soviet natural gas was under 3 percent in 1970 and despite an 80 percent increase in Azeri gas extraction, the country's relative share of the national total changed very little during that time (see Table 1.1).

Azerbaijan was never known for its natural gas potential, but it could regard its oil sector with considerable pride, although output was now slowly declining, as emphasis first was shifted to the Ural-Volga region following the end of World War II and then to Western Siberia upon the discovery of the giant Samotlor oil field in the middle 1960s.

TABLE 1.1
Central Asia and Azerbaijan's Share of Soviet Crude Oil and Natural Gas, 1970 (Percentage of Total)

	Crude Oil	Natural Gas
Central Asia	8.2	24.3
Azerbaijan	5.7	2.8

Source: CIA, National Foreign Assessment Center, *International Energy Biweekly Statistical Review*, ER IEBSR 78-003, February 8, 1978, pp. 24, 26.

In 1970, Azerbaijan could supply just under 6 percent of Soviet oil production, which by now had risen to 7.06 million b/d, through the rapid development of oil fields in the Ural-Volga region. The supply from these oil fields reached a plateau in the mid-1970s and was replaced by supply from even more prolific oil fields in Western Siberia. West Siberian oil grew at astonishing annual rates, jumping from 630,000 b/d in 1970 to 2.96 million b/d in 1975. In the interim, oil production in Azerbaijan continued its slow but steady decline, from 400,000 b/d in 1970 to 340,000 b/d by 1975, depressing its relative share of the Soviet total oil output that year to under 3.5 percent. By that time, Azerbaijan was much more important to Moscow for its oil equipment manufacturing industry than for the crude oil and natural gas it produced.

Just as Soviet oil production had become heavily concentrated in Western Siberia, so did the natural gas sector shift its geographic center, to north western Siberia. To illustrate this, in 1977, Western Siberian natural gas (together with small volumes originating in the Far East) accounted for just under 21 percent of the national total. Ten years later, in 1987, this share had risen to slightly more than 62 percent. In the interim, although quite reasonable volumetric growth could be demonstrated, the importance of Central Asian natural gas declined from 32 percent to less than 19 percent. By 1990, the last full year before the dissolution of the Soviet Union, the role of Azerbaijan and Central Asia, in terms of domestic crude oil and natural gas supply, was as shown in Table 1.2.

However, by this time, another trend had set in, a much more worrisome one for Soviet officials. Soviet oil production was in difficulty, a

TABLE 1.2
Central Asia and Azerbaijan's Share of Soviet Crude Oil and Natural Gas, 1990 (Percentage of Total)

Crude Oil	Million Barrels per Day (mbpd)	Percentage Share of National Total
Azerbaijan	0.250	2.2
Kazakhstan	0.516	4.5
Turkmenistan	0.112	0.1
Natural Gas	**Billion Cubic Meters (bcm)**	**Percentage Share of National Total**
Azerbaijan	9.9	1.2
Kazakhstan	6.8	less than 1
Turkmenistan	87.3	10.7

Source: Robert E. Ebel, *Energy Choices in Russia*, Center for Strategic and International Studies, Washington, DC, 1994 and *Energy Choices in the Near Abroad*, (CSIS) 1997.

peak of 12.5 mbpd had been reached and a dramatic decline had set in around mid-1988. There were no more easy finds in Western Siberia to develop. What could be done? These officials saw the future of Soviet oil in Azerbaijan, in Kazakhstan, and in the Kara and Barents seas. But with the dissolution of the Soviet Union in 1991, that prospect was taken away from them. The oil potential of Azerbaijan and Kazakhstan now belonged to these newly independent nations. Of course, Russia still had the Kara and Barents seas, plus considerable onshore potential, but it lacked the capital and the technology to develop this remaining potential by itself. Indeed, it was the lack of capital that had brought about the production decline and the breakup only magnified that weakness.

Azerbaijan and Kazakhstan were quick to recognize that their political and economic future would be largely based on the power of oil and lacking any capability to exploit that power, looked to the financial, technological and managerial strengths of the major international oil companies to help them secure this future. With the encouragement and support of their respective presidents, Azerbaijan and Kazakhstan moved relatively quickly to provide the kind of investment environment required. Turkmenistan did so to a lesser extent.

Russia, conversely, judged that it did not need any outside assistance to help the oil sector regain its past level. After all, just a few years ago Russia was leading the world in oil production and, if sufficient capital could be made available, the production decline could be replaced by a growth curve. Investors, however, wanted more than to be just a source of capital. The foreign oil companies sought access to oil through production-sharing arrangements and through joint ventures. Expressions of nationalism intruded however, and foreign investors despite showing willingness, have been disappointed in the Russian response to date.

This is not to say that investors in Azerbaijan, Kazakhstan and in Turkmenistan have enjoyed unencumbered access. They have not, but both sides have persevered and have found mutually advantageous solutions to those problems which inevitably arise in any country where the oil industry is predominant.

As noted above, Turkmenistan was a major contributor to the Soviet natural gas sector, with extraction peaking in 1990, the year before the Soviet breakup, at 87.8 bcm. Since that time, extraction has declined precipitously, largely because Russia has denied Turkmenistan access to its pipeline system. Without that access, production has had to be shut in. Turkmenistan's dependence on Russian largesse carries a strong political message for that country.

Kazakhstan's oil sector was doing quite well during the years immediately preceding independence, with production levels steadily increasing from 262,000 b/d in 1970 to 516,000 b/d in 1990. However, after peaking in 1991 at 532,000 b/d, production went into decline as investment fell and the giant Tengiz oil field was slow in coming into play, with its growth in output not sufficient to offset field declines elsewhere.

Unfortunately, while the end of the Cold War changed a number of things, it could not change the physical isolation of Azerbaijan and Central Asia from desirable Western markets. To sell in these markets, pipelines would have to be built across international boundaries and would have to transit regions torn by civil and ethnic strife, partly latent, partly open conflict. Under normal conditions, pipeline construction and the routes chosen are a matter of economics. Not so in this part of the world. The issue of pipeline construction very quickly acquired a political flavor and became widely known, whether correctly or not, as the "Great Game," a designation readily adopted by the media.

Perhaps it is a game but if so, it is a game without an end. Operating pipelines as well as oil and gas flows will be enduring magnets, attracting nations and individuals alike. There is much at stake here, well beyond the scope of simple additions to world crude oil and natural gas supply, and all the players know it. The Russian newspaper *Izvestiya* understood what was going on when it stated that "control over pipelines will be the most important factor of geopolitical influence in the Transcaucasus and Central Asia in the next century."

One can hope that the completion and successful operation of export pipelines to carry Caspian crude oil and natural gas – construction of which must be justified economically – will help solidify the political and financial independence of the exporting countries. Unfortunately, attractive alternatives to the wise use of export-derived income stand in the way, in the form of corruption, investments for show and the temptation of an arms race. Few exporting countries have been able to resist using the power of oil incomes for self-aggrandizement.

In the introduction to his book *The Russian Rockefellers*, Robert W. Tolf describes in considerable detail what life was like in Baku around the turn of the century:

> In this primitive land a new style of Western dishonesty was combined with the ancient Eastern to produce a new and shrewder commerce which did not honor or even recognize the concepts of a later age. It was said that among the several hundred Baku oil barons, only ten were honest: an Armenian, a Swede, and eight Mohammedans.

Today's temptations, not surprisingly, mirror those of the past.

When oil companies search for crude oil and natural gas, three completely different kinds of risk are confronted.

- *Geological risk:* Are the underground rock formations such that they would trap and hold crude oil or natural gas?
- *Technical risk:* If crude oil or natural gas is discovered, is available technology capable of bringing these resources to the surface, at a cost which supports project viability?
- *Political risk:* The higher the political risk, the higher the return must be. Can the project yield a financial return commensurate with the perceived political risk?

As a prospective investor considers a venture in the Caspian Basin, how would these three risks be evaluated? In reality, much depends upon the senior management of the prospective investor. Is the investing company risk-averse or is it willing to take up the challenge of the Caspian Sea? Is the investor looking for a quick return or is he there for the long run? For the most part, investors in the Caspian Sea are not risk-averse and look forward to a long and mutually beneficial presence.

Media coverage of developments in the Caspian Sea has been intense and sometimes misleading, as the search for eye-catching headlines often leaves a false impression with the reader. In sum, the world has not found in Caspian oil a substitute for Middle East oil, as some writers would have us believe.

Then why invest in the Caspian at all? The response is simple. Oil companies must continually search for new supplies to replace those barrels now being produced and there is nothing as attractive as something which has been denied but which is now available. Second, the Caspian oil potential is world-class. Third, this potential can be developed within an acceptable time-frame only through the involvement of multinational oil companies. And fourth, and possibly most important, the bulk of the new oil to be developed will be for the export market; local demand is comparatively small and is likely to stay that way.

Nations are prisoners of geography, and no one nation enjoys in unlimited fashion all of the fruits that geography can bestow. Some, by accident of nature, are rich in energy resources but totally lacking in other strengths. Some are dynamic in all of the virtues we may respect but poor in natural resources. This makes for a shrinking and increasingly interdependent world. At the same time it also makes for conflicts among nations, as each seeks to maximize strengths and minimize weaknesses, while preserving and hopefully enhancing its stature among its peers.

It is out of this conflict that the issues of today and tomorrow emerge for Azerbaijan and Central Asia. In the following chapters, the authors explore the many aspects of Caspian oil, ranging from contentious export pipeline issues to the impact of Caspian oil on Arab Gulf oil producers and on OPEC member-states, and the relevance of Caspian oil for the world. An answer to the question as to what has brought foreign investors to the Caspian in the first place is provided by a discourse on the attraction of Caspian oil reserves. Equally important, the legal environment is thoroughly explored, as are regional politics and political trends.

In underscoring the difficulty ahead in developing Caspian oil Laurent Ruseckas emphasizes that while the region seems relatively calm today, the potential for new instability exists. Pipeline route selection has dominated the scene both for potential transit countries and the Western oil companies, complicated in part by the prospect that an Iranian route may in time become possible.

Paul Gregory points out that no major projects can go forward without a reasonably stable legal regime and that includes resolution of the dispute over whether the Caspian is a sea or a lake. Geoffrey Kemp expands on the larger issue of regional politics, concluding that all five Caspian littoral states have serious domestic problems and that their political stability is uncertain. Moreover, the struggle for control of Caspian oil and natural gas holds far-reaching implications for regional security and could eventually impact on the stability of the Arab Gulf.

Building on Dr Kemp's analysis, Shirin Akiner provides a risk assessment of Azerbaijan, Kazakhstan and Turkmenistan, discussing in detail the threats to their stability. These threats include political succession, clan and regional rivalries, increasing poverty levels, and corruption. In a wider regional content, the Azerbaijan–Armenia conflict, religious fundamentalism and the drug trade, among other factors, also contribute to instability.

Julia Nanay answers the common question: why are Western oil and gas companies so attracted to the Caspian Basin? The answer is found largely in the lack of investment opportunities worldwide. But how will successful development of Caspian oil impact on the Organization of Petroleum Exporting Countries (OPEC) and on the world market? OPEC has been careful in its assessment of Caspian oil as a potential rival. Wilfrid Kohl, in his review of Caspian oil development, notes that Caspian oil will be an important addition to non-OPEC oil supply and will postpone somewhat that point in time when OPEC oil production once again exceeds non-OPEC output.

Others view the future somewhat differently. Amy Myers Jaffe, with her conservative outlook for Caspian oil, given the reality of Caspian Basin development, suggests that the Arab producers' market control may be relatively unaffected over the short to medium term. Fadhil Chalabi expands on the question of the Caspian and its impact on Gulf oil in his discourse, but quickly concludes that Caspian oil holds little prospect of offering any serious competition to Gulf oil. However, he does caution that Caspian oil may eventually threaten Gulf oil unless the region's countries adopt more market-oriented economic investment and price policies than has been the case.

Will Caspian oil hold any particular relevance for the world energy market? Sarah A. Emerson looks beyond pipeline politics and addresses not only that relevance but also the challenges that Caspian oil will present to OPEC.

Whether by chance or by foresight, the author Louis Fischer came close some 73 years ago to capturing the sense of what is taking place today in and around the Caspian Sea. In his book *Oil Imperialism*, he wrote:

> It will be sad to see how the magnet of oil draws great armies to the Caucasus; it will be fascinating to examine how the oil companies mobilize their forces of diplomacy to fight their battles across green tables and behind the scenes . . .; it should be enlightening to study how far the foreign policies of nations, in the matter of recognition, of credits, etc. are influenced by that universal lubricant and irritant – oil . . .

Today the multinational oil companies *have* returned to Baku in search of oil to meet the requirements of tomorrow. Battles *are* being fought over green tables and behind the scenes. Foreign policies *have been adjusted* so as to ensure that national interests, in part conducted through the oil companies, at all times are being served.

In sum, not all that much has changed.

---

* Special thanks are due to Herro K. Mustafa and Mary Abraham of the Emirates Center for Strategic Studies and Research for their editorial role in this volume.

1

Caspian Oil Development: An Overview

Laurent Ruseckas

Just ten years ago, the areas on either side of the Caspian Sea – Central Asia to its east and the Caucasus to its west – were largely unknown. These regions were provinces of the Soviet empire, important to the outside world neither politically nor economically. Today, things have changed around the Caspian. Eight obscure new states gained their independence suddenly in 1991 with the collapse of the Soviet Union. Since that time, some of these new countries have attracted an increasing amount of attention from governments, the private sector, international policy circles and the media.[1]

This rise to relative prominence by a few of the Caspian states – primarily Azerbaijan, Kazakhstan and Turkmenistan – is due largely to oil and gas. For a handful of international energy companies searching for new areas in which to explore and produce hydrocarbons, the Caspian first appeared on the radar screen during the final years of the Soviet period. Interest grew steadily from 1991 to 1997, and this period saw the region emerge as the world's largest new area for upstream oil investment.[2] This commercial interest in Caspian oil has had a notable effect on the attitudes of external states toward the region. For many states – the United States prominent among them – oil has given the Caspian region a political and economic significance that it would not otherwise have enjoyed.[3]

At the same time, there is much more to the strategic importance of the region than just oil. The Caspian Sea lies at the crossroads between Russia, Turkey, Iran and China. With the Soviet Union's collapse, the region's newly independent states (NIS) find themselves dominated by none of these outside powers – but linked to them in diverse ways by geography, economics, politics and culture. These neighboring countries perceive both opportunities and risks in the Caspian, and each has come to see fundamental national interests at stake in the region. The geopolitical game that has emerged, as external states try to gain influence in the Caspian region while guarding against any possible threats from the region to their own security, has become inextricably linked with oil development, and in particular with the issue of pipelines.

For oil companies operating in the Caspian, the pipeline issue has been fundamental from the start. The region's oil-producing countries are landlocked, and existing pipelines will not suffice to bring projected new volumes of Caspian oil to market. For the companies, financing and building expensive, transborder pipeline infrastructure represents the region's greatest challenge – one that is political as well as commercial. From the perspective of the relevant governments, pipelines can have weighty political implications as they will help establish the economic orientation of the Caspian region, and perhaps its political orientation as well, in the twenty-first century.[4]

Just as the hype about the "Caspian oil rush" was nearing its peak in late 1997, the price of oil began to decline. The sustained downturn in the world oil market that lasted through the spring of 1999 has noticeably shifted the mood of the budding Caspian oil industry, from euphoria to a more balanced realism. The need for new pipelines is not only the core of the region's geopolitical maneuvering; it is also the single largest impediment to the economics of developing its oil. Unlike most of the oil produced elsewhere, against which the Caspian must compete, new volumes of oil from Azerbaijan or Kazakhstan are expected to involve transportation costs of between US$2.50 and US$5 per barrel before it has even reached a sea terminal. Given production costs that are not among the world's lowest, the problem of transportation costs calls into question the competitiveness of Caspian oil supplies.[5]

This chapter attempts to provide an overview of some of the key political and commercial issues surrounding Caspian oil development. In particular, it emphasizes the pipeline question, which will largely determine (along with geology and oil price) the impact of Caspian oil on world markets, and thus its ultimate importance for the Arab Gulf.

Geological Realities: How Much Oil Does the Caspian Region Hold?

Any discussion of the future role of the Caspian region in the world oil market must begin with the geological realities beneath the ground. Unfortunately, this is a topic about which there has been a great deal of confusion – although less within the oil industry itself than within media and foreign-policy circles in the United States, Europe and elsewhere.

The oil reserves of the Caspian region – defined here as Azerbaijan, Kazakhstan and Turkmenistan – which can be considered proven, amount to 25 to 30 billion barrels.[6] These discovered volumes alone would make the Caspian comparable to the North Sea as an oil province. Two fields, Tengiz in Kazakhstan and the Azeri-Chirag-Güneshli (ACG) megastructure in Azerbaijan, are world-class giants, each with reserves in the range of five billion barrels (and probably more for Tengiz). Particularly given the fact that these fields were discovered in the Soviet period, thus eliminating geological risk as a factor, the opportunity represented by Tengiz and ACG for the first international oil companies to move into the Caspian was enormous.

However, a perception took hold during the euphoric phase of the Caspian boom that the region's reserves amounted to far more than the North Sea. The figure that has been often cited in the press for total Caspian oil reserves is 200 billion barrels – a huge number that has given rise to comparisons between the Caspian region and the Gulf. These comparisons, and the figure of 200 billion barrels on which they are based, are quite simply not justified. The chief source for this estimate appears to have been a report delivered to the United States Congress by the Department of State in April 1997.[7] This provided a rather conservative assessment of proven Caspian oil reserves, but cited an optimistic figure of 200 billion barrels as the possible reserves of the region. Among journalists at large, however, the difference between "proven" and "possible" reserves is not well understood, and many journalists and their readers ended up with the impression that the Caspian is another Gulf.

Of course, there *is* a huge upside potential in the Caspian region. The offshore area is mostly unexplored, and the potential for future large discoveries is thought to be great. There are high hopes for an exploration well that is currently being drilled at Kazakhstan's Kashagan prospect by a large consortium of European, Japanese and American companies. If Kashagan is a success, it could conceivably add another 10 billion barrels to the region's total reserve base. Estimates of 40 to 60 billion barrels as the ultimate reserve base of the Caspian region are judged to be reasonable by most geologists familiar with the region.

Still, the Caspian is destined to play a supporting role rather than a leading one in supplying the world oil market in the future. It is unlikely that the region will at any point account for more than 4 to 5 percent of world oil production – and even to reach such a figure will require the satisfactory resolution of a number of logistical, commercial and political difficulties.

The Transportation Challenge

The major obstacle to the development and export of Caspian oil is the lack of transportation infrastructure available to link future oil production from the region to its future markets. The Caspian basin is landlocked, and the pipeline system left behind by the Soviet Union is capable of exporting only limited volumes of oil from the region – and that through Russia, which has proven to be highly problematic as a transit country. Therefore, development of new export routes has been a top priority for companies and their host governments alike.[8]

However, movement toward transportation solutions has thus far been slow, and this has consequently slowed the overall pace of development in the Caspian. At the root of the problem is the fact that transborder pipelines sit squarely at the intersection of politics and economics. In order for a multi-billion-dollar pipeline project to succeed, it must have strong political support from the states involved. Indeed, this support must be strong enough to offset the political opposition from external states that is so often present. At the same time, a pipeline must have the commercially grounded support of the relevant commercial entities. This group consists primarily of the companies that will be producing and shipping the oil in question, but it also includes potential project lenders or equity investors.

To achieve this broad support requires a close alignment of interests among several governments and companies. But particularly as projects have moved beyond their early planning stages and closer to the moments of truth when investment decisions must be made, it has become clear that aligning these interests is exceedingly difficult.

For both host countries and potential transit countries, prospective pipelines are seen to have a profound political as well as economic significance. The question of pipeline routes has evolved into the centerpiece of the international tug-of-war for influence in the Caspian region which some have taken to calling the "new Great Game" in a reference to the nineteenth-century competition between Britain and Russia for influence in Central Asia. From the perspective of a producing state, pipelines are much more than just physical infrastructure – they are also political infrastructure. A pipeline promises to solidify a country's geopolitical links with (some of) its neighbors as well as the oil-consuming world at large, while also guaranteeing the oil revenues that will hopefully bring domestic peace, security and prosperity.

But from the commercial perspective of an investor, the political importance of a pipeline is a side issue. For a company hoping to produce oil in the Caspian, the purpose of a pipeline is to allow transport of oil in an appropriate time-frame while minimizing transportation costs. For a commercial lender or equity stakeholder, participation in a pipeline project must be justified on economic terms – keeping in mind that the expected rate of return must be adjusted upward to reflect perceived risks, which are still high. Considering the financial weakness of the new states of the Caspian region, pipeline projects are unlikely to move forward without meeting these commercial criteria – no matter how strong their political logic may seem to host governments or interested external states.

The Pipeline Situation: Today and Tomorrow

Kazakhstan, which currently produces about 650,000 b/d, can export some portion of that through one existing pipeline that links Kazakhstan's oil-producing area to the Russian pipeline system. However, capacity in this line is limited, and much of this oil is diverted to Russian refineries, that pay less than the world-market price. The main project to build a new pipeline from Kazakhstan to a new port on Russia's Black Sea coast – the Caspian Pipeline Consortium (CPC) venture – has been moving forward only in fits and starts since it was first conceived in 1992. The reasons for this are too long to enumerate here, but they are primarily commercial rather than political. However, the final obstacles to beginning construction appeared to have been cleared away in November 1998, and since that time the project has been moving forward, with contractors being chosen and initial work beginning. The current schedule calls for this pipeline to become operational in late 2001. The CPC pipeline will more than double Kazakhstan's export capacity at first, and eventually quadruple it.

Other pipeline options to export Kazakhstani oil are being explored for the more distant future: a pipeline to China; one through Afghanistan to a Pakistani port; several alternatives through Iran; and a trans-Caspian pipeline that would link up with whatever export route is chosen for Azerbaijani oil. Three of these routes (all but the Afghan variant) have been tested by producers who have exported small batches of oil – at high cost – via tanker and/or railroad. And particularly if large new oil reserves are discovered offshore, at least one of them will be needed by 2010. Yet each of these pipelines promises to be both expensive and politically controversial.

From Azerbaijan, which had no capacity for pipeline exports until late 1997, the Azerbaijan International Operating Company (AIOC) consortium began in 1998 to export small volumes of oil – the initial ceiling for this route is 100,000 b/d – through a refurbished pipeline that heads through Russia (including Chechnya) to the Black Sea. Azerbaijan's experience with the use of this pipeline thus far has been mixed. On one hand, the Russian pipeline monopoly Transneft has, for the first time in its existence, signed a long-term contract with a fixed tariff, and then lived up to the commercial terms. Yet the pipeline was plagued by numerous shut-offs and operational problems throughout 1998 due to the situation in Chechnya, culminating in the indefinite shutdown of the line in the spring of 1999. While Russia continues to propose that Azerbaijan's use of this pipeline be greatly expanded in the years ahead, the new round of intense military conflict in Chechnya would seem to make this an improbable route for the foreseeable future.

In early 1999, AIOC completed a second pipeline for "early oil"[9] that transits Georgia, terminating at the new port of Supsa from which the first cargo was lifted in April 1999. This route offers lower tariffs than the Russian early-oil pipeline, and for that reason it had become Azerbaijan's favored export route even before the shutdown of the Russian pipeline made it the country's only near-term export route. However, this is a 20-inch pipeline and therefore not a long-term solution: its initial throughput capacity is 100,000 b/d, with potential for expansion to perhaps 200,000 b/d but not much more. Thus AIOC and the Azerbaijani government are considering what to do with later, larger volumes of oil. The construction of large new pipelines through Russia, Georgia and Turkey has been considered. Meanwhile, producers in Azerbaijan are cautiously exploring or thinking about export options involving Iran – alternatives which could have substantial economic merit but which are strongly opposed by the US government as part of its broader policy of containment and isolation of the Islamist regime in Tehran.

The political desires of Washington, Ankara and most importantly Baku have led to the selection of a route linking Azerbaijan with the Turkish port of Ceyhan on the Mediterranean Sea. In Washington's vision of an "east–west corridor" for Caspian oil and gas flows, a trans-Caspian oil pipeline would feed into the Baku–Ceyhan line, turning this into a huge export system which would meet most of the region's needs while bringing commercial benefit through economies of scale.[10] A pipeline to Ceyhan would also have clear environmental benefits in comparison to routes terminating on the Black Sea, which would increase the traffic of oil tankers through the sensitive Turkish Straits, the Bosphorus and Dardanelles. In the long run, most observers agree that some pipeline bypassing the Bosphorus must emerge – either a major pipeline to the Mediterranean Sea or the Arab Gulf, or perhaps a short bypass pipeline linking the Black Sea to the Aegean or Adriatic seas. However, the relative dearth of proven reserves in Azerbaijan's offshore sector – beyond AIOC no other foreign consortium in Azerbaijan has found sizeable proven reserves of oil – the project appears to be commercially problematic for the time being.

Whatever routes are chosen, it is clear that transportation costs associated with bringing Caspian oil to market will be unusually high by world standards. Given that average production costs in the region are relatively high, at least in comparison with the Gulf, the Caspian will not be a source of low-cost oil. The oil-price environment that prevailed in 1998 and early 1999 served to bring this issue into sharper focus. No companies have backed away from the Caspian (excepting two consortia in Azerbaijan that were unsuccessful in their exploration plans), but their eagerness to move forward quickly with massive pipeline investments has been somewhat tempered.

Iran: A Promising Market

With transportation costs looming so large, the possibility of selling Caspian oil in less distant markets using infrastructure that will not require billions of dollars in investment is an interesting one for producers in the region. The most promising market in this category is northern Iran. The proximity of northern Iran's refineries – primarily those in Tehran and Tabriz – make Iran's future role in Caspian oil transportation one of the major questions for the future. Iran has proposed the establishment of swap arrangements, wherein Caspian crude could be delivered to these northern refineries in exchange for cargo or Iranian crude lifted from Kharg Island.[11] But thus far, US policy has successfully deterred companies from pursuing Iranian pipeline options in earnest.

Today the situation may be beginning to change, particularly in the wake of the waiver to the Iran–Libya Sanctions Act which the US government granted to Total and its partners for the South Pars project.[12] Many Western, non-US companies are hoping that Iranian routes can soon be cautiously pursued. The construction of a major, dedicated pipeline from the Caspian region to the Gulf via Iran is a possibility for the long term, but it is far from being a sure thing. The removal of the political constraint posed by the policy of the American government would only be the first step toward the realization of such a project. But swaps into northern Iran, first from Turkmenistan and later from Azerbaijan and Kazakhstan, are very likely to emerge as a significant mechanism for the marketing of Caspian crude.

Sources of Political Risk

Transportation is a political problem as well as an economic one. Export routes in any direction must transit countries whose stability is not guaranteed. Political risk must therefore be a major concern for producers operating in the region. Today, with the conspicuous exception of Chechnya, the region seems relatively calm, especially when compared to the chaos and conflict of 1992–95. But most of the region's conflicts have not been satisfactorily resolved, and the potential for new instability exists.

Some of the major sources of political risk for Caspian oil production and export are as follows:

- *Nagorno-Karabakh conflict:* Azerbaijan fought and lost a war with Armenian forces between 1991 and 1994 over Nagorno-Karabakh, an autonomous region within Azerbaijan's borders which is now under Armenian control. Although a cease-fire has been in effect for more than five years, the basic issues that gave rise to the conflict have not been resolved. Peacekeeping efforts led by the United States, Russia and France have made little tangible progress despite concerted negotiations throughout much of 1999. With Armenia feeling increasingly threatened as the "odd man out" in the Caspian oil boom, the possibility that this conflict will reignite cannot be ignored.
- *Chechnya Conflict and Russian Instability:* The resumption of military conflict between the Russian military and the Chechen separatists has highlighted the fundamental unsteadiness of the entire former Soviet space. The post-Soviet revolution is still underway, most visibly in Russia, whose people and political elites have not yet decided what sort of state, economy and foreign policy they want for Russia. States like Azerbaijan and Kazakhstan are no longer dependent completely on Russia, but until Russia is reasonably stable, it will pose a major source of uncertainty for the Caspian region. In particular, the war in Chechnya today threatens to undermine the hard-won stability of Georgia, the crucial transportation hub for the region.
- *Domestic politics in the Caucasus:* While the two key states of the South Caucasus – Azerbaijan and Georgia – have stabilized considerably since their first years of independence, they are not yet in the clear. Stability in these countries rests to a large extent on the foundation of their two leaders, Presidents Haider Aliyev and Eduard Shevardnadze. But succession scenarios for both states are hazy; the passing of these two leaders, when it occurs, will usher in a dangerous and uncertain period.
- *General geopolitical rivalry:* The lines of engagement in the geo-political struggle for influence in the Caspian region have tended

to foster the formation of two opposing blocs, with Turkey, Azerbaijan, Georgia and the United States on one side; and Russia, Iran and Armenia on the other. (The Central Asian states would fall somewhere in between.) The hardening of these countries into two clear blocs is still some distance away, and there are many political and economic forces working to prevent it. But to the extent that this "blocification" of regional politics continues, it will make pipeline projects which require broad regional cooperation more difficult.

It should be stressed that political risks of the sort described here are nothing new and they have thus far proven manageable. Compared to the situation in 1993, when the political obstacles to Caspian oil development seemed insurmountable, today's achievements in the Caspian represent a huge triumph over uncertainty and risk. Furthermore, the potential for all sides to benefit from the development of Caspian oil provides a powerful incentive for cooperation and regional partnership. Doom and gloom are not justified; but at the same time, neither should it be expected that the region's politics will offer a smooth, trouble-free environment for the years ahead.

Conclusion

From today's vantage-point, it appears that a combination of difficult economics and high political risk may push Caspian oil development onto a path quite different than the one that was envisioned as recently as 1997. Much of the energy and effort devoted thus far to solving the Caspian pipeline question has been focused on a handful of multi-billion dollar megaprojects. But a strong case can be made that these large projects will materialize slowly – and some not at all.

The most important external factor affecting Caspian oil development will be, of course, the price of oil. In the market conditions of the second half of 1998 and early 1999 – with the spot price of a barrel of Brent crude in the US$10 to US$13 range – Caspian oil development is a difficult proposition. Sustained prices in this range would not spell the end for the Caspian, but it would certainly set a threshold that not all upstream projects in the region would meet. On the other hand, a structural shift in the oil market which pushed Brent prices into the US$8 to US$10 range or even lower would mean serious trouble for the Caspian as a hydrocarbon region.

Of course, the runup in oil prices that began in the spring of 1999 has made the concerns that loomed so large less than a year ago seem somewhat outdated. Still the psychological holdover of low oil prices continues to affect the perspectives of the oil companies active in the Caspian region. Companies are working hard to ensure that the massive investments being contemplated will be profitable even in a relatively low-price environment. Given the transportation costs and the growing conventional wisdom that many discoveries in the Caspian offshore area are likely to contain not only liquid hydrocarbons but also significant quantities of difficult-to-market gas, it may be difficult for medium sized discoveries to meet commercial thresholds.

Caspian oil development has gained a great deal of political and commercial momentum since the first foreign companies came there at the end of the 1980s. At this stage, we can be certain that it will move forward in some form. Particularly if oil prices remain at high levels, and assuming some level of exploration success in the Caspian offshore, it is even possible that the more optimistic projections will be borne out: the Caspian region could conceivably be producing between three and four mbpd by 2010. However, this scenario seems less plausible than it once did. The future is more likely to bring a Caspian oil boom that is more modest, and more gradual.

2

Developing Caspian Energy Reserves: The Legal Environment

Paul R. Gregory

Need for a Stable Legal Regime

The "Caspian" denotes the area of substantial hydrocarbon reserves located in countries bordering or near the Caspian Sea. The Caspian "does not mean just the sea anymore, but rather a region defined by oil."[1] The ownership of onshore reserves in the Caspian region is not subject to serious question, insofar as post-Soviet national boundaries have been largely accepted. With substantial undeveloped reserves located under the Caspian Sea, the substantive legal and political issues relate to the ownership of the Caspian Sea's seabed. Similarly complex legal and political issues complicate the creation of a transportation infrastructure for Caspian energy resources.

Current oil and gas resources of the Caspian region have been compared to oil reserves of the North Sea and gas reserves of North America. Potential oil reserves of the region may be comparable to Saudi Arabia. The estimates for combined reserves of Azerbaijan, Kazakhstan and Turkmenistan range from a low of 15 to a high of 200 billion barrels of oil and a low of 230 to a high of 650 trillion cubic feet (tcf) of gas.[2] The wide range of these estimates (2 to 17 percent of world oil reserves and 5 to 12 percent of world gas reserves) shows that it is still not known how rich these Caspian resources are. The larger estimates include potential reserves and may be tainted by the "optimism" of officials in producing countries. Also, if potential reserves in other parts of the world are included, the share of the Caspian states is much lower than 17 percent and 12 percent of oil and gas respectively. Most of the possible reserves (about 160 billion barrels) are under the Caspian Sea.

There are no reliable estimates of the reserves of Russia and Iran near or under the Caspian Sea, but they may potentially be significant.[3] However, even if the more conservative estimates are taken, there can be no doubt regarding the immense strategic implications of the Caspian region.

The Caspian region requires a stable legal regime to develop, produce, transport and market its hydrocarbon reserves. Property rights to the reserves must be defined and accepted; its oil and gas industries must be subject to stable and reasonable tax regimes. Commercial law must define corporate structures and management practices. Transportation rights must be recognized. Because the Soviet system primarily developed "easy" oil reserves, Caspian reserves yet to be developed require massive injections of long-term capital, which must be obtained in international capital markets. International capital will not be forthcoming without a legal regime strong enough to withstand potential domestic political opposition to foreign participation in, and ownership of, mineral resources. The legal status of the landlocked sea is not just an academic or judicial problem.[4] It has become a major business problem for the 40-some oil companies such as Chevron, Mobil (now Exxon Mobil), BP, Atlantic Richfield and others involved in the region.

This chapter addresses the creation of a new legal regime in the Caspian region following the collapse of the Soviet Union in December of 1991. In particular, I consider:

- The process of defining legal property rights to offshore hydrocarbon reserves under the Caspian Sea either by comprehensive agreement or bilateral negotiation.
- The application of international legal arrangements to transport and market Caspian oil and gas and to provide for environmental protection.
- The creation of national regimes of commercial, tax, subsoil use and environmental laws and regulations concerning the development of hydrocarbon reserves, their production, taxation and marketing.

Outstanding domestic, multilateral and international legal issues must be resolved to create the legal stability that attracts long-term investment. Experts estimate that US$50 billion or more of investment dollars are required to develop the region's energy reserves.[5]

The Caspian Sea is bordered by several states.[6] It is the largest inland body of water in the world, comparable in size, depth and salinity to large semi-enclosed marine seas, such as the Great Lakes separating the United States and Canada. The Caspian is not connected directly to the open sea and is completely surrounded by the land masses of four former Soviet republics – Russia, Kazakhstan, Azerbaijan and Turkmenistan – and by Iran to the South. The "Caspian" can be defined more broadly than these five littoral states to include, in all, the ten states (both former Soviet and other states) in close proximity to the Caspian Sea that would be involved not only in oil and gas production but also its transportation.[7]

Geopolitics and the Legal Regime

In the Caspian region, it is not possible to divorce political from legal issues, especially with regard to international legal arrangements.[8] Legal issues can be resolved only if the various parties desire to find solutions. The interplay of politics and legal matters is complicated by dynamic changes and by the conflicting interests of the different players.

Russia has been torn between its desire to use the Caspian to strengthen its political interests in the Commonwealth of Independent States (CIS) and the need to preserve its monopoly over export pipelines versus the desire of "private" Russian companies, such as Gazprom or Lukoil, to use Caspian opportunities to maximize their commercial interests.[9] With Russia's conclusion of the war in Chechnya and the growing strength of Russia's multinational energy concerns, Russia has demonstrated an increasing willingness to seek real solutions to the Caspian legal regime.

Newly independent Azerbaijan, Kazakhstan and Turkmenistan view the commercial development of Caspian energy resources as vital to their economic development and independence. Revenues from new oil and gas will help them resolve political and military problems (such as Azerbaijan's conflicts with Armenia) and maintain their undemocratic political systems. These three former Soviet republics must balance their commercial interests with the need to co-exist with their neighbors. Kazakhstan has maintained that "all Caspian littoral states should be allowed to take full advantage of the resources of the sea" and that national zones of ownership should not infringe on existing cooperative agreements concluded with international oil companies. Azerbaijan has relied on international partnerships, balanced geographically, to promote energy development. Turkmenistan, which has outstanding disputes with Azerbaijan over ownership of Caspian deposits, has supported the vague formula that "all Caspian problems should be resolved between the littoral states," while lately (1998) concluding agreements with international partners to develop its offshore reserves.[10] Turkmenistan's dispute with Azerbaijan over ownership of offshore deposits has caused it, at times, to favor a regional settlement, fearing that bilateral negotiations may not provide it with a proportional share of Caspian wealth.

Iran, as a major oil producer but a potential transporter of Caspian oil, wishes to be involved in regional decision-making and to minimize the prospect of Azeri separatism, engendered by the large number of Azeris living in northern Iran. Iran, also motivated by anti-Americanism, may wish to inflict harm on Western interests by playing a spoiler role.[11] But Iran must also worry about being excluded from major Caspian development and transportation projects. Iran has found itself increasingly isolated in recent years. Iran's promotion of joint resolution of the Caspian legal regime (discussed below) may ultimately prove a pretext to slow Caspian development and to compensate for what may be its small share of Caspian undersea wealth. Moreover, Iran's position will be affected by future changes in its relationship with the United States.

In the Caspian region, stated reasons for actions often conceal hidden agendas and cannot be taken at face value. Russia's opposition to an undersea pipeline, couched in terms of fears of environmental damage, conceals its opposition to a non-Russian oil transportation route. Similarly, Iran's calls for a common environmental regime may conceal a hidden agenda of forcing a more favorable division of undersea resources.

Countries outside the region, such as China, Pakistan, the Gulf states, the United States, industrialized Asia and Europe view Caspian energy resources each according to their interests: Consumers, such as China, Turkey and Europe, welcome new energy production in world markets, while producers, such as the Arab Gulf states (who are not participating directly in the process) fear a weakening of world energy markets. Countries outside the region influence the legal environment indirectly or through international conventions. For example, both the United States and Western Europe have promoted Western-style legal and political regimes for the area.

International Law: Undersea Reserves

The former Soviet republics of the Caspian region must create new legal rules of the game to replace those of the era of Soviet administrative planning. They must work out their own national legal systems as well as multinational regional arrangements. International law provides precedents for dealing with undersea property rights and associated legal issues, but the eventual application to the Caspian of international legal conventions for oceans, seas and lakes depends upon the economic and political interests of participating countries. International law can provide a convenient starting-point, but it will not resolve what are basically political and economic issues.

The United Nations (UN) Law of the Sea focuses on freedom of navigation, not a key issue in the Caspian region, where offshore underground resources and their exploitation are the issue. Traditionally, "internal waters," such as rivers and lakes, are not subject to the law of the sea.[12] Because the Caspian Sea is landlocked, navigation rights of third parties are not at issue – only the navigation rights of the bordering states themselves need to be considered. It is easier to liberalize navigation rights when there is no military threat; hence much discussion has focused on the Caspian Sea as a "demilitarized zone of peace."[13]

International practice has customarily partitioned inland bodies of water surrounded by more than one nation among the parties, although partitioning can be combined with joint management and control (a condominium) of some activities. Opportunities for mixed solutions abound: The seabed can be partitioned, while the rights to the water itself or environmental regimes may be jointly managed. In examining the historical record, Oxman concludes: "If there is any lesson to be learned from the modern law of the sea, it is that there is a vast middle ground between these options [partitioning versus condominium] depending upon history, geography and other circumstances."[14] International law does not bind the Caspian to any particular solution – partition or condominium. Precedent appears to favor partition of mineral rights, but, at a minimum, the bordering states are free to select their own legal-regime variant.

The reality is that most lakes and semi-enclosed seas bordered by more than one state, such as Lake Constanz or the Great Lakes, have been partitioned.[15] Partitioning has been accompanied by varying degrees of cooperation and joint management relating to transit rights, fishing, underground cables and environmental protection.[16] With respect to underground resources, practice shows no significant difference between lakes and marine semi-enclosed seas. In lakes, as in other internal waters and the territorial sea, the sovereignty of the surrounding states typically includes the subsoil.[17]

Partitioning of seabed hydrocarbon and mineral deposits (assuming communications and environmental concerns can be accommodated) is supported by overwhelming state practice. This is true even in the Mediterranean Sea, where most coastal states have partitioned shelf jurisdiction over the seabed and subsoil, but have not partitioned the water mass itself beyond the territorial sea.

Regional versus Bilateral Agreements

Two approaches have been advanced to resolve the Caspian's legal status following the collapse of the Soviet Union: a comprehensive regional settlement arrived at unanimously among the surrounding states, or bilateral agreements among the various parties negotiated according to their self-interest. The history of the post-Soviet Caspian legal regime shows the failure of the initial intent to achieve an all-inclusive regional agreement. Instead, the Caspian countries have entered into bilateral agreements to divide the Caspian basin into national sectors, especially over the past two years.

Soviet Period Agreements

The breakup of the Soviet Union necessitated the creation of a new legal regime for the Caspian Sea. Until the dissolution of the Union of Soviet Socialist Republics (USSR), the only two Caspian littoral states were Iran and the USSR, but no Soviet–Iranian boundary in the Caspian Sea was ever delimited.[18] A Soviet–Iranian treaty of 1940 gave each party an "exclusive right of fishing in its coastal waters up to a limit of 10 nautical miles" and stated that the "parties hold the Caspian to belong to Iran and to the Soviet Union." These and other treaties provided for free and equal navigation by ships of Iran and the USSR throughout the Caspian and that no third state had any rights in the sea, including the right of navigation.

The USSR carried out limited activities of exploration and exploitation of mineral resources under the Caspian Sea, primarily off Azerbaijan, more than 10 miles from shore. The USSR never offered a share of the proceeds to Iran and Iran never protested the USSR's development efforts. In fact, Iran conducted its own unilateral activities in seabeds adjacent to its own borders.

The USSR did not delineate formal Caspian boundaries among the republics. However, in practice, oil and gas assets and operations (largely onshore) were assigned to the various republican ministries of oil and gas. The delineations of reserves to republican ministries formed the basis for the post-Soviet distribution of oil and gas production (see Table 2.1). Onshore reserves, assigned by this Soviet practice, have not been contested by the various parties. Agreements on the regional distribution of Soviet assets and liabilities – aviation assets, pipelines, foreign debts etc. – were a part of the founding of the CIS in 1991 in which the Caspian states of the FSU participated. Since the dissolution of the Soviet Union, these *de facto* boundaries have been respected by the successor states with relatively few outstanding disputes (Azerbaijan and Armenia in Nagorno-Karabakh, Georgia versus Abkhazian separatists and Azerbaijan and Turkmenistan disputes over an offshore Caspian field).

TABLE 2.1
Energy Production of the Former Soviet Republics, 1997

	Crude Oil (Million Tons)	Natural Gas (Billion Cubic Meters)
Russia	306	570
Azerbaijan	9	6
Kazakhstan	26	5
Turkmenistan	5	35

Source: DIW DataBank.

During the last years of the Soviet Union, as Gorbachev sought to attract Western investment to the USSR, international tenders for deposits in republics were handled jointly by the Soviet government and by the government of the republic.[19] Since the breakup of the USSR, negotiations have been conducted exclusively by the governments of the newly independent Caspian states, despite some challenges to their authority by Russia and Iran.

A Regional Solution Based on Unanimity

During the Soviet period, the regulation of the Caspian legally rested with the two Caspian littoral states: the USSR and Iran. The Soviet–Iranian Treaties of 1921 and 1940 did not address the ownership and utilization of underground resources located beneath the Caspian Sea, but informal practice, especially on the Soviet side, created an informal delineation of the Caspian seabed into national territories.[20]

Regional Regimes: CASCO and ECO

The gathering of the Caspian states in Tehran in February 1992 marked the first regional step toward creating a comprehensive Caspian legal regime. In Tehran, Iran proposed the creation of the Caspian Sea Co-operation Organization, (CASCO) to provide a venue for periodic meetings to resolve issues of shipping, fishing and maritime resources.[21] At the subsequent Almaty Meeting of foreign ministers in May 1995, it was decided that the legal departments of the respective CASCO foreign ministries would be in charge of resolving the legal regime and a meeting of legal experts, held in Tehran in June 1995, concluded that the Caspian's legal status should be decided by unanimity. At a meeting in Ashgabat in November of 1996, the foreign ministers of the five Caspian nations reiterated (at least this was Iran's interpretation) that the legal status of the Caspian Sea should be decided unanimously by the five littoral states.[22] In effect, CASCO was to be (at least in Iran's view) the primary instrument for negotiating the Caspian Sea condominium agreement.

The Caspian states have also joined and constitute a part of the ten-nation Economic Co-operation Organization, (ECO), which now comprises six former Soviet republics in Central Asia, Afghanistan, Iran, Pakistan and Turkey. ECO was brought into being with the signing of the Treaty of Izmir in 1977 and brings together 325 million people from ten non-Arab Islamic countries.[23] The foreign ministers of the ECO countries meet periodically to deal with political and economic matters concerning the region, including the development of Caspian oil reserves and its transportation.

The Iranian Position

The Iranians (and also initially Russia) intended CASCO to be used to promote a new joint legal regime in support of their position, at the time, that the Caspian Sea should be jointly managed by the five littoral states. In this regard, the Iranian legal position[24] was based upon the 1921 and 1940 Soviet–Iranian Treaties, the agreement of CIS countries to honor the treaties of the USSR (Almaty Declaration, December 1991) and the CASCO Agreements from Tehran, Almaty and Ashgabat. The Iranian case for joint management (condominium) was also based upon the perceived need for joint management of the Caspian Sea's fragile ecosystem, which is particularly sensitive to oil and gas exploration; and the need to contain the arms race and to control illegal trafficking. According to Iran:

> The Caspian – whether construed as a sea, a semi-closed sea or a lake – encompasses a vast body of riches that can meet the requirements of the region over the short, medium and long run. It is only through the expression of good will, cooperation, integration and collaboration of the littoral states that optimal results can be gained.[25]

Brice Clagett has supplied an extensive legal argument against the Iranian position for condominium: "Initially, it may be noted that, while certain writers have discussed the theoretical possibility of common ownership of lakes and inland seas, no such condominium appears even to have existed." He further notes that a leading scholar (Verzije), after an exhaustive study, concluded that examples of common ownership "do not to my knowledge actually exist." He rejects the Iranian claim based on 1921 and 1940 Soviet–Iranian treaties granting common fishing rights outside of the 10-mile limit by noting that fisheries are often governed by regimes that bear no relation to the regime governing seabed and subsoil resources. Clagett also argues that even if the language of the treaties could be interpreted to mean that the seabed and subsoil resources of the Caspian are owned in common, the "common owners" would be Iran and the USSR. None of the USSR successor states would award 50 percent of all the resources of the Caspian to Iran, without providing any guidance as to how the remaining 50 percent is to be divided amongst themselves. Moreover, the littoral states can deal with pollution matters through a separate treaty, which would not be inconsistent with each littoral state's right to ownership of the seabed and subsoil resources.[26] Clagett further maintains that no practical difficulties would result from the delimitation of Caspian boundaries, at least for the purpose of exclusive rights to seabed and subsoil resources. Similarly, there is ample international legal precedent on methods of delimitation.[27]

Divergence of Positions

Ultimately, the viability of a comprehensive regional approach depends upon a reasonable degree of common interests. As noted above, the Caspian Sea littoral states have divergent goals and objectives and these divergences have widened over the years. The lack of progress on a formal unanimous agreement has become increasingly apparent. Those Caspian states with rich national sectors and in need of massive capital injections have less interest in a common settlement. By May 1998, the Caspian states were far apart on issues of pipelines, Western investment and other matters. In the words of the Kazakh foreign minister, "We must be frank in saying that our positions regarding the Caspian Sea have substantially parted."[28] Table 2.2 shows the strong incentives to proceed unilaterally. Each country in the region has substantial oil or gas reserves which are waiting to be developed. These states fear being left behind by others and a loss of revenues if resources are commonly owned. Iran's future "national" sector may be among the poorest, although geological work remains to be done.

TABLE 2.2
Distribution of Reserves under the Caspian Sea

	Oil: Billion Tons (Billion Barrels)	Gas (Trillion Cubic Meters)
Russia	1.0 (7.3)	1.0
Kazakhstan	3.0 (21.9)	1.5
Turkmenistan	0.5 (3.7)	1.0
Azerbaijan	2.5 (18.3)	1.5

Source: International Centre for Petroleum Business Studies, Moscow, Reuters, April 17, 1998.

Bilateral Agreements

After the Soviet Union's breakup, Kazakhstan and Azerbaijan – and later Turkmenistan – began to negotiate agreements with international companies to develop independently their offshore hydrocarbon reserves without waiting for a regional solution. Kazakhstan's agreement with Chevron concerning the Tengiz field was concluded simultaneously with the end of the Soviet Union. Azerbaijan's formation of an international consortium to develop its offshore Caspian reserves provided the impetus for the first open conflict.[29] These early negotiations were initially protested by Russia, which claimed that the newly independent republics were still subject to USSR laws and that the utilization of Caspian marine resources would have to be resolved unanimously and jointly. Russia represented that the Caspian was an inland lake; therefore, the condominium principle should apply. In October 1994, Russia supported its position before the UN with the claim that the Caspian Sea had no connection to the open seas and should be treated as a jointly managed lake.

Iran's and Russia's initial efforts to promote unanimous regional agreements were frustrated by the other Caspian littoral states' pursuit of their own commercial interests. Azerbaijan's agreements with international partners to jointly develop its offshore reserves through the Azerbaijan International Operating Company, (AIOC) were motivated by the Azeri need for national economic development and foreign capital. In the Azeri view, for example, "only real economic development may help achieve real independence."[30] Turkmenistan's need to develop and transport its natural gas reserves independently of the former Soviet transportation system caused it to look to foreign partners. Turkmenistan, despite its disputes with Azerbaijan over ownership rights,[31] concluded in the summer of 1998 that it had to move forward in its own sectors to avoid being left behind.[32]

Economic self-interest provided the driving force for bilateral agreements delineating national sectors of the Caspian seabed. It was understood by the parties that international investment would not be forthcoming unless legal property rights were established – a process that could not be completed quickly under a condominium agreement. Moreover, joint operation would threaten to dilute the revenues of countries holding particularly rich deposits. Azerbaijan began the process by asserting its ownership to offshore fields that had been allotted to its republican energy ministry during the Soviet era. In this regard, Azerbaijan's claims were largely accepted by the other littoral states, except for a dispute with Turkmenistan over the status of one offshore field.[33]

In the latter part of the 1990s, Russia's position began to change in favor of partition. At the end of 1996, Russia proposed that the coastal zone should be expanded from 10 to 45 miles and that only beyond this zone should underground resources be jointly developed. Russia then entered into negotiations with Kazakhstan to partition national zones in the northern Caspian. In the northern Caspian region, bilateral negotiations between Russia and Kazakhstan and Russia and Azerbaijan (conducted largely in 1998) concluded in agreements based on the principle of the division of the seabed into national zones, while maintaining a uniform legal regime for environmental protection (including the laying of pipelines on the Caspian seabed) to prevent "utter ecological and economic banditry."[34]

The Russian and Kazakhstan agreement of July 6, 1998 mapped out boundaries of the seabed but left the water itself in common hands[35] and called for the resolution of fishing and environmental issues by the five Caspian countries working together. The agreement signed by the Presidents of Russia and Kazakhstan divided the seabed of the northern Caspian along a median line equidistant from each country's shores, effectively opening the way for both to develop their energy reserves independently. The agreement includes another important provision – that Russia and Kazakhstan cooperate on oil and gas pipelines which are crucial to getting the Caspian's huge reserves out of this landlocked part of the world.

Twelve international oil companies, including BP, Statoil AS, Total SA and Amoco Corp and Unocal, are now involved in consortia for developing offshore Caspian fields near Azerbaijan, Turkmenistan and Kazakhstan. The growing involvement of Russian companies, interested in Caspian oil deposits outside the Russian zone, has pushed the Russian government to change its stance on the legal status of the sea.[36]

Iran continues to protest bilateral agreements that divide the Caspian seabed into national zones, as is indicated by the following statement of the Iranian Ministry of Foreign Affairs:[37]

> . . . The Islamic Republic of Iran resolutely opposes any unilateral measures adopted by the Azerbaijan Republic and reiterates that any steps in this direction contradict the aforementioned agreements that are binding on Azerbaijan as a legal successor of the USSR according to the regular norms of international law and the Almaty

> Declaration of December 21, 1991 . . . The full responsibility for the consequences of such unlawful measures and steps, including damages inflicted on other littoral states, lies with the states that violate the legal status of the Caspian Sea.[38]

Further, on May 4, 1998, Iran's Foreign Minister repeated his country's position that it would oppose any unilateral exploitation of the Caspian Sea's oil reserves:

> Any understanding on principles which would eventually lead to a unilateral exploitation of the Caspian Sea . . . would be unacceptable to the littoral states, particularly Iran and would be rejected . . . Iran declares its readiness for talks based on goodwill . . . towards a just and reasonable settlement which would reform and complete the legal status of the Caspian Sea and its future exploitation.[39]

Iran has also protested the proposed construction of a cross-Caspian undersea pipeline, saying that it would be "a violation of the rights of other littoral states."[40] Tehran's position in favor of a condominium has been left with no supporters and Iran's foreign ministry is faced with growing opposition from the country's oil ministry.[41]

Environmental Regimes

Even if conclusive agreements on ownership rights were concluded, regimes for environmental protection and for transportation would still have to be established. The UN Convention on the Law of the Sea contains a comprehensive and powerful environmental protection regime that could be applied to the Caspian Sea.[42] In a multi-state lake divided into national zones, like the Caspian, issues of transboundary pollution must be addressed by the legal regime. Pollution of a lake, even a large lake, is likely to affect fishing and other interests of the other bordering states. In an undivided lake subject to joint management, externality issues could be dealt with by joint management. Hence, from a theoretical point of view, there are strong arguments in favor of a joint management regime for environmental issues.

The environmental provisions of the draft articles of the International Law Commission on the law of the non-navigational uses of international watercourses, which include lakes, were substantially influenced by the UN Convention on the Law of the Sea.[43] Article 7 of the draft articles provides, "Watercourse states shall exercise due diligence to utilize an international watercourse in such a way as not to cause significant harm to other watercourse states." Article 20 provides: "Watercourse states shall, individually or jointly, protect and preserve the ecosystems of international water courses." Like Part XII of the Convention, the draft articles contain detailed provisions on water pollution, exotic species, cooperative management, notice and consultation and other matters. Russia, in particular, has stressed the need for strong environmental regulation, particularly in view of the importance of Caspian fishing to the Russian economy.[44]

Environmental protection of the Caspian Sea requires, first, agreement on the legal environmental regime and second, an effective means of monitoring and enforcement in an area not known for its observance of laws. The offshore Caspian Sea is an area in which substantial environmental damage could occur, due to offshore drilling accidents, pipeline breakages and seismic problems.

One overlooked advantage of the participation of international concerns in the exploitation of Caspian offshore reserves is their experience in developing offshore reserves in an environmentally safe manner and their natural caution stemming from fear of substantial liability claims by damaged parties. Such caution may be less evident among national companies operating in the region, which are more accustomed to Soviet-style attitudes toward environmental damages.

Despite expressions of concern by littoral Caspian states, comprehensive agreement on the regime of environmental protection is still lacking and discussions have not progressed much beyond talking points. Russia, in particular, has been vocal in expressing concern about the uncoordinated use of Caspian reserves. The planned construction of deep-water oil pipelines, the Russians claim, is especially alarming due to active seismic zones, underwater geysers and the moving seabed.[45] Russia has expressed its concerns on the likelihood of ecological disaster associated with laying pipelines on the Caspian seabed.[46] But the commercial stakes are high and the Caspian states wish to receive revenues as soon as possible. The most likely scenario is the development of offshore fields well before any consistent environmental regime is put in place.

There can be little dispute that the Caspian Sea shows signs of an impending ecological disaster.[47] Not only is much of the Caspian Sea polluted, but the oil leaking from offshore wells may have prevented evaporation and may have caused the waters of the Caspian Sea to rise. All countries bordering the Caspian Sea are affected from an economic point of view, with the commercial interests of the Russian fishing industry the most seriously affected.

The need to protect the Caspian's ecosystem is a fact accepted by all parties. Despite this consensus, it will prove difficult to put in place a regional system that is adequately monitored and enforced, given the rush to be the first to develop offshore fields.

Transportation Issues: Legal Regimes

The landlocked nature of the Caspian Sea makes the resolution of the transportation regime essential. Expensive-to-develop hydrocarbon resources, no matter how rich, cannot be exploited without transportation facilities to world markets. Because none of the former Soviet states in the Caspian region has direct access to world markets (except Russia), transportation systems must cross one or more national borders. For economic reasons alone, transit countries are in a position to impose tariffs that alter the costs and risks of developing the primary resources.[48] Transit countries, moreover, may use transit rights for political rather than economic gain. Resolving the Caspian's transportation regime, therefore, is a key element to exploiting Caspian reserves.

The Caspian states (with the notable exception of Russia) are signatories of the Energy Charter Treaty (ECT), which imposes a particular legal regime on its members. In December 1991, fifty-one states (the Central and East European countries, the European Union and its members, the states of the former Soviet Union, the United States and other non-European members of the Organisation for Economic Co-operation and Development (OECD), such as Japan, Canada and Australia) agreed on the European Energy Charter.[49] The European Charter's multiple objectives include the promotion of energy cooperation between countries formerly divided by the Iron Curtain, assisting transition to a market economy, maximizing the efficiency of production, conversion, transport, distribution and use of energy, minimizing environmental problems and ensuring the security of energy supplies. Obviously energy supply security and transportation security are interrelated.

The signatories undertook to pursue the Charter's objectives and principles and to negotiate legally binding agreements. After three years of negotiations, the ECT, together with the Protocol on Energy Efficiency and Related Environmental Aspects, was opened for signature in December 1994. The ECT has since been signed by all signatories of the European Energy Charter with the significant exceptions of the United States and Russia.

Azerbaijan, Kazakhstan and Turkmenistan, as signatories of the ECT, could in theory apply its framework to deal with some of their outstanding issues, such as environmental protection and transportation issues.[50] The Energy Charter Treaty addresses directly the issue of energy transportation as a key element in the security of energy supplies. The ECT ensures that:

- energy and energy products can transit through third countries
- countries must take the measures necessary to facilitate such transit
- countries must not interrupt the flow of energy in the event of any dispute over terms and conditions of transit.

If these provisions were faithfully observed in the Caspian region, the ECT should provide a suitable legal framework for transportation of Caspian energy to markets. Divergence of political interests and of public and private economic interests threaten observance of ECT rules by countries involved in the Caspian transportation chain. Russia, as the current near-monopoly transporter of Kazak oil and Azerbaijan "early oil," may not wish to lose the leverage that its monopoly position provides.[51] Moreover, the private interests of Transneft, the Russian monopoly oil pipeline corporation, may differ from the public interests of Russia in this matter. As a private company, Transneft would be expected to act as a monopoly rent-seeker and do everything in its power to preserve its monopoly rights. Russia has long used its control of oil and gas pipelines to achieve political leverage over other former Soviet republics.

Much has been written on the economics and geopolitics of the Caspian's transportation alternatives. The outcome of the struggle over transportation routes could change the balance of power and has rendered the region a "great geopolitical game." Iran wishes Caspian oil to transit its territory. China wants Caspian oil for its own economic development. Russia wants to retain its transportation monopoly and so on.

Recent actions show the lack of intent to obey the transit rules of the ECT. The CPC has already indicated that consortium members will receive first priority on the pipeline to Novorossisk – an important consideration given the CPC pipeline's quite limited capacity. Moreover, the pipeline's operator, Transneft, has a long history of discriminatory practices.[52]

ECO provides one possible venue for decision-making concerning alternate transportation choices. In fact, much discussion within ECO in recent years has focused on transportation issues, with general resolutions that transportation choices should be based upon economics rather than politics.[53] A confounding issue is that US companies and their affiliates are not permitted by Executive Order to have business dealings with Iran, which tends to rule out their involvement in transportation projects that cross Iran's territories. Although US executives hope that Washington will back away from its insistence that Caspian oil and gas not be transported through Iran, the US State Department has declared that Washington would not exempt Caspian pipelines from sanctions.[54]

Currently, commissions comprised of government officials from the participating countries or ECO meetings are used to formulate transportation proposals and to discuss options.[55] The key stumbling-block is that the various ECO members back different projects (Turkey supports a route through Turkey, Iran through Iran) and thus it is difficult for ECO members to reach a common position.[56]

Local conflicts in the near-Caspian and transCaucasian zones make the onerous task of developing and transporting Caspian oil even more complicated. The war in Chechnya seriously compromised the very idea of exclusive reliance of Caspian oil-producing states on Russian processing and transportation and precipitated the search for alternatives. Likewise, the Nagorno-Karabakh conflict has affected Azerbaijan's efforts to attract foreign investment. Despite this handicap, Azerbaijan has been the most successful recipient of oil-related foreign investment in the region. Alternate routes through Afghanistan are threatened by continuing political instability in this region and Georgia has just achieved a degree of political stability.

Although all Caspian energy-producing states remain dependent on Russian pipelines or on alternatives yet to be constructed, they vary in the difficulty of solving transportation constraints. Some oil and gas is currently being transported through Russia but subject to serious restrictions and uncertainties. According to Shenoy and Gulen:

> Clearly, Azerbaijan and Kazakhstan are the countries that will desperately need additional export route capacity. Currently, pipeline capacity allowed by Russia to Azerbaijan, Kazakhstan and Turkmenistan is limited mainly to eight mt of Tengiz oil sent to the Baltic Sea through Druzbha pipeline and one mt of Azeri oil sent to Novorossiysk. Alternative transportation methods of trains and barges, or swaps with Iran are cumbersome but provide a solution to the problem in the shortrun . . . Although Turkmenistan's exports will remain a fraction of those by Azerbaijan and Kazakhstan, the country will still need to find out a route out for its oil . . . Turkmenistan is probably in the worst shape. Most pipeline routes will have to go through several potential competitors' territory. The most direct route through Iran is not workable at the present time.[57]

National Legal Regimes[58]

The Caspian's development depends not only on resolution of regional or international legal arrangements, but also on successes in creating stable *national* legal regimes. Even if all regional (transnational) issues were resolved satisfactorily, deficient national legal regimes alone could prevent the development of Caspian reserves. Clearly, long-term investments require a more stable legal and political environment than short-term investments. Caspian countries, as signatories of the ECT, have obligations to protect foreign investment, but the more important foreign investment guarantees are imbedded in their national legislation.[59] Foreign investment is required because of the technological requirements and capital costs of developing the geologically difficult resources of the region.[60]

Alternate Models of National Legal Regimes

The Soviet oil and gas industry was managed in a highly centralized manner in the era of the administrative-command economy.[61] The

Ministry of Oil Industry and Ministry of Gas Industry set production quotas, established delivery plans and drew up plans for investment and renovation. Oil and gas were produced by regional production associations that were under the supervision of the ministry. The various departments (*glavki*) of the Ministry managed different aspects of production, transportation and sales. Directors of production associations were judged on the basis of quantity plan targets, such as tons of oil produced, volumes of oil or gas transported, gallons of gasoline refined, or numbers of oil wells drilled. When Soviet oil and gas was sold to the West, domestic producers simply received the low domestic price while the difference went into state revenues.

Corporate control was exercised by Soviet state committees and ministries. These handed down assignments to managers, who were expected to fulfill the plan. Managers, given the complexity of the plan, could exercise a degree of managerial autonomy, but, by and large, the system's objective function was the fulfillment of physical production targets with less emphasis on cost reductions, efficiency and technological change. The Soviet system created a management philosophy geared toward physical outputs with relatively little attention to "shareholder" interests – with the state, or society, as the shareholder. A common phrase from the Soviet period was "Property belongs to everyone and hence to no one."[62]

The national legal regimes of Russia and other former Soviet states bordering the Caspian Sea have had to be created anew with the collapse of the USSR.[63] These legal regimes could be patterned after a number of legal models; choices include the Anglo-Saxon model of privately owned companies, regulated by independent regulatory commissions, pushing the participants toward more competitive outcomes; the German model of privately owned companies wherein competition is largely excluded by public franchises and the transportation system does not provide open access; or the French Model, which is based on integrated, largely state-owned monopolies operated according to directives from central government ministries. The legal regime must decide whether energy companies will be state or privately owned, whether integrated companies (perhaps directed by state authorities) will be encouraged and which regulatory regime will be adopted.

The resolution of these institutional processes will determine whether the Caspian economies are to operate on the basis of "rule of law" or "negotiation and adhocery." Western businesses seeking to enter the Caspian market have had to face this issue squarely, when confronted with the choice of case-by-case contracts approved by high political authorities or arrangements based on the impersonal rule of law. The Caspian countries themselves have had to decide whether they can develop their energy infrastructures based on *ad hoc* arrangements with political authorities, or whether to focus on developing the legal rules of the game.

The Russian Legal System

Among the former Soviet republics bordering the Caspian Sea, Russia has made the most serious efforts to develop a rule of law for oil and gas resources. The Russian Constitution of December 1993 and the earlier Law on Underground Resources (1992) divide statutory responsibility for the Russian energy industry between the federal government and regional governments on issues such as land ownership, subsoil use and natural resources generally.[64] The Russian legal regime has been influenced by the UK and US models. Russia's laws and regulations spell out licensing procedures, production-sharing arrangements and tax provisions.

The management and ownership of the Russian oil and gas industry have undergone significant change during the transition. The Russian successor organizations to the former ministries, the Ministry of Fuel and Energy, which supervises all fuel and energy resources (oil, gas, coal and electricity), the Committee for Geology and the Committee for Mineral Resources, exercise weaker control than during Soviet times, through licensing, quotas, influence on legislation and some regulation of prices and rates. Economic power has largely shifted to the privatized or corporatized Russian vertically integrated oil and gas companies. The former Soviet gas ministry is now a joint stock company, Gazprom, which retains a monopoly over the production, sale and distribution of natural gas, playing on public fears of disruption of gas supplies and its massive political clout to maintain its monopoly. Oil transportation remains dominated by a monopoly company (Transneft) and natural gas transportation remains under the control of Gazprom. Russian oil is now produced by a number of vertically integrated oil companies; there is no single oil company (as in other Caspian states) that represents the state, although the state has retained ownership in most companies.

The major challenges to creating Russia's legal system have been the lack of political consensus between the executive and legislative branches and problems of enforcement: An ongoing battle has been waged between the legislative and executive branches over tax reform, production sharing and privatization. The Russian Constitution's strong presidential system makes it possible, at least temporarily, to run the Russian economy via presidential decrees, but government by presidential decree does not create the long-term stability that attracts foreign and domestic capital, especially with a weak and ailing President. It will be difficult to attract foreign capital, to pass meaningful tax reform and finalize production-sharing regulations until the legislature participates more fully in decision making. The enforcement of laws and regulations has also been deficient; the lack of observance of laws and regulations has been widespread. Specifically, Russian citizens, both private and juridical, have lacked access to a strong court system that can enforce compliance and arbitrary bureaucratic actions continue.

Legislative stalemate and an inability to reach binding decisions complicate the Russian business environment. Foreign companies had to wait years for parliamentary passage of production sharing laws,[65] only to learn of additional regulatory hurdles imposed by a reluctant parliament. Although all parties agree that Russia's tax system is untenable, it is not possible to gain agreement on tax reform. Unresolved issues of federalism expose companies to a complex web of national and local taxes.[66]

Kazakhstan

In its early years of independence, Kazakhstan used its early pilot investment project (the Tengiz field with Chevron) as a model legal regime largely devoid of official legislation and regulation. Initially, Kazakhstan operated projects, approved by presidential decree, through the ministries of oil and gas. Starting in late 1994, Kazakhstan began the process of upgrading its legislative and regulatory framework for the oil and gas industry, passing its Law on Foreign Investment in December 1994 and its Oil Law in June 1995.[67] In December 1996, Kazakhstan passed a new, improved tax code, which is viewed as partially complete at best. The creation of a rule of law was deemed essential; the earlier system required presidential decrees for virtually all actions, large and small.

Kazakhstan moved to corporatize its oil industry in March of 1997 with the formation of KazakOil, a state-owned company that replaced the Ministry of Oil Industry. KazakOil is a true national company closely entwined with Kazakhstan's top leadership. The new company, patterned after the State Oil Company of the Azerbaijan Republic (SOCAR), was designed to act as a gatekeeper and participate in the country's energy projects. KazakOil has already penetrated the Caspian Pipeline Consortium.[68] The vice president of KazakOil is married to President Nazarbayev's daughter. As is characteristic of the other Caspian states (except Russia), KazakOil controls virtually all of Kazakhstan's oil reserves[69] and typically serves as the state's representative in international consortia. KazakOil's dominant role has been threatened at times, with the privatization of certain fields, which were sold in part to foreign investors, but the fear of foreign ownership has prompted the Kazak government to continue to transfer additional energy assets to KazakOil.

For offshore development, Kazakhstan has used Production Sharing Agreements (PSAs) with international consortia partners.

Azerbaijan

Throughout much of its independence period, Azerbaijan ignored Western legal models and relied instead on the *ad hoc* approach of considering each investment project individually via the sponsorship of high-level Azeri officials. Azerbaijan has focused on attracting foreign investment and has sought a geographic balance of foreign partners. The early passage of the Law on Foreign Investment (January 1992) reveals Azerbaijan's priority interest in attracting foreign investment to develop its energy infrastructure.[70]

Azerbaijan is a civil-law codified country; its new constitution was approved later than in other former Soviet republics (Referendum of late 1995), although it did pass a Law on Foreign Investment relatively early (January 1992). On the whole, Azerbaijan's laws are general framework laws which take on meaning only through implementation of laws and decrees, often in the form of presidential decrees. Decrees are usually published in newspapers.[71] Cases are handled at the highest political level, including case-by-case approvals and parliamentary actions. The Azeri parliament ratifies major PSA contracts, thereby solidifying their legal status. Azerbaijan's interests in major projects are represented by SOCAR, which typically serves to represent state interests and is often an obligatory partner in major projects. Foreign partners, hence, are often confronted with arbitrary changes that are dictated either by President Aliyev or by SOCAR.[72]

Turkmenistan

In Turkmenistan, the oil and gas industry has been directly subordinated to the President throughout Turkmenistan's period of independence. There are no forces that challenge President Niyazov, the self-proclaimed "father of all Turkmens." In 1992, Niyazov created his own democratic party, which anyone desiring a political career must join – a throwback to the Soviet Communist Party. Niyazov has directed policy and a parliament, although naturally opposed to reform, has simply rubber-stamped Niyazov's proposals.

Turkmenistan, like Azerbaijan and Kazakhstan, has lately come to realize that some rule of law is required to attract foreign investment and has sought to create a belated legislative framework. In December 1996, Niyazov signed the Law on Hydrocarbon Resources and in April 1998 he signed a partial privatization decree. In May of 1997, Niyazov endorsed model contracts that offer favorable treatment to subsoil users operating under PSA agreements. Prior to these laws, arrangements with international partners were handled primarily through presidential decrees. The Hydrocarbon Law established a tax regime, streamlined the procedure for issuing licenses and set out tender procedures. Its drawbacks, however, include Turkmenistan's limited liability in case of dispute and uncertainty concerning the ability of Turkmen officials to review contract terms.[73]

Turkmenistan has been pressured by European governments to pass political and economic reforms that would eventually transfer power from the president to parliament.[74] In the Turkmen parliament, however, there is little enthusiasm for reform. Parliament has had little interest in creating a cohesive economic and legal structure. In other words, the president seems more inclined to favor progressive legislation than parliament. As in Azerbaijan and Kazakhstan, parliament tends to oppose liberalization and privatization. Members of parliament consider the energy industry to be a significant political lever. In discussions of the political economy of reform, reform opponents are many. The President is the only possible advocate of legal reform.[75]

After three years of hesitation, Turkmenistan approved an international tender for the exploration and development of its offshore hydrocarbon reserves. The tender was initiated in the form of an Ordinance signed by President Niyazov, which established a new commission to conduct the tender. Turkmenistan is fearful of being left behind in the development of its Caspian resources by procrastination.[76]

Common Features

In all the former Soviet republics bordering the Caspian, the struggle over the legal regime has been between those who favor reliance on *ad hoc* personal arrangements and those who advocate an impersonal rule of law. No Caspian state has yet established an effective rule of law. With the slight exception of Russia, they are run by dictatorships in which the president has complete appointive powers; the most influential political figures are relatives of the dictator; and the press is managed by the dictator. In effect, the FSU Caspian states are run in a manner not very different from the Soviet past. In its most extreme form (such as in Turkmenistan), people who oppose the president are declared mentally ill.

Given the competition in the region for scarce investment capital, each Caspian country has come to realize that it must eventually offer a rule of law in order to attract foreign investment. Despite the strong support for energy investment by the respective presidents, sponsorship by the highest official in the land may not be enough. After some early experiences in which host countries changed the terms of contracts with foreign investors, it was recognized that foreign investors will divert their capital to other countries if treated arbitrarily. Although it has proven difficult to establish stable rules of the game, especially for taxation, these countries realize that they will get no significant investment if they do not.

The trend has been in the direction of allowing foreign investors more authority, such as in Kazakhstan and in Turkmenistan, where it is realized that without foreign involvement, ambitious development plans cannot be realized. All Caspian countries realize the importance of eventually shifting from investment policies molded by individual decisions of high-ranking officials to government by legislation. In all cases, there is an understanding that foreign investors will come only if there is stability in the legal environment. To quote a Turkmen official, "Our goal is for Turkmenistan's oil and gas regulations . . . to eventually match international standards."[77]

No Caspian country has yet to succeed in establishing a satisfactory "rule of law." Foreign investors are still subject to arbitrary tax regimes, erratic support by sponsoring officials and parliamentary opposition. Corruption is rampant in all the Caspian littoral states; the Caspian states now serve as an important transit area for worldwide drug trafficking.[78] Given the growing competition for foreign investment, foreign investors can "vote with their feet" and punish national governments by shifting their investment dollars to other areas. The countries of the region (other than Iran) realize that resolution of these legal issues is required because, as one fund advisor notes, "The international investment community will direct its funds towards those areas where the risks are least."[79]

Conclusions

The Caspian is indeed a "great prize," which will not be won without considerable effort and risk taking. Clearly, long-term infrastructure investment requires a reasonable "rule of law" and a consensus on the legal "rules of the game." Participants on both sides of the equation – international investors and the countries themselves – find themselves in a form of competition, similar to the famous "Prisoners' Dilemma" of economics. The countries risk being left behind by other countries that offer superior legal rules of the game. International investors, displeased by the lack of a rule of law, cannot withdraw from the game for fear that competitors will remain in the game and win.

The Caspian region can look to international legal precedents to create its legal regime. International precedents in other parts of the world demonstrate that joint management of complex undersea mineral resources is rare, if not nonexistent. Moreover, international legal precedents show that ownership issues can be handled differently from other matters such as environmental regulation and fishing rights. Joint ownership and management of Caspian undersea resources (the Iranian position) is clearly losing the battle. The pattern of bilateral agreements for the northern zone of the Caspian will eventually extend to the southern Caspian, where Azerbaijan, Turkmenistan and Iran must eventually reach agreements on national seabed zones.

So far none of the international organizations – ECO, CASCO, the European Energy Charter – has proven to be an effective venue for resolving environmental, transportation or ownership issues in the Caspian. Although the European Energy Charter binds its signatories to strict rules concerning oil and gas transportation, there is no evidence that these rules are being obeyed.

The Caspian Sea is inexorably moving towards partition into national sectors because of the fundamental pressure of the profit motive. Those countries that delay, will risk falling behind and perhaps sacrificing claims to other countries. Russia's growing support for partition is attributable to this realization, plus the desire of its own "private" oil companies to participate in the Caspian. Russia's support for partition has been bought in part through participation of its companies and the prospect of transportation revenues. Russia has come to the conclusion that, considering its other interests in the Caspian, the country is better served by a sectoral division of at least the sub-surface resources. Russia intends to preserve its economic and political interests through the involvement of its "private" companies and through pipelines from the Caspian running through its territory.

Iran is the solitary holdout for a condominium of joint ownership of the resources, but a softening of its position may soon be seen. "Iran is keen that at least some of the Caspian oil should go south through its territory to the (Middle East) Gulf," according to a senior energy analyst at the Centre for Global Energy Studies in London.[80] There could be room for negotiation between Kazakhstan and Iran and between Turkmenistan and Iran that, in return for supporting the division of the Caspian, pipelines will be built through Iran.

Official positions tend to conceal hidden agendas. Russia uses environmental arguments to oppose an undersea pipeline that would compete with Russian pipelines. Iran similarly uses ecological arguments to promote joint management of the Caspian's resources and is moved by both political considerations and by the fact that, under general partitioning rules, its share of offshore Caspian resources would be small.

The *de facto* partitioning that has occurred (and accelerated in 1998) has resolved the seven-year dispute of whether the Caspian Sea is a sea or a lake. Only Iran continues to argue the point. Given that there is more consensus on the need to preserve the Caspian Sea's ecosystem, progress is expected in the direction of joint management. However, there is no effective multi-state regional body that can deal with this issue.

The three most serious obstacles to the development of Caspian Sea energy resources are first, the tumultuous row over transportation routes complicated by high geopolitical stakes and by widely divergent interests of the parties, second, the slow progress in establishing reasonable "rules of law" in each of the Caspian states and third, the unresolved issue of who will be in charge of developing and managing Caspian resources. On the issue of the rule of law, there appears to be a common pattern: first, each state has attempted to attract foreign investment via *ad hoc* arrangements with high-level political officials, usually the president of the country. They then discover that foreign investors cannot be satisfied indefinitely with presidential decrees and they move slowly toward a legislative framework, which has been, most often, opposed by a parliament that dislikes foreign ownership and participation. Moreover, the lack of a rule of law means arbitrary actions and unpredictable enforcement on the part of political officials and administrators.

An effective rule of law requires an effective means of enforcement. In no Caspian littoral state is there an effective legal system to protect domestic and foreign investors. Russia's financial collapse of late 1998 is likely to weaken the rule of law in Russia. Foreign investors in other Caspian states are still poorly protected by the local legal system.[81] Insofar as the major infrastructure projects of the region are being developed by consortia of international, Russian and local companies, another major unresolved issue is whether these consortia will be operated as profit-maximizing entities or will be subject to local management and local rules.[82]

The *de facto* division of the Caspian Sea into national zones may eventually require a comprehensive settlement. The huge international investments required to develop the offshore Caspian will be deterred by territorial ownership disputes, such as between Turkmenistan and Azerbaijan and, to a lesser extent, by disputes between Russia and Azerbaijan. Consider the disputed offshore field called Serdar by the Turkmens and Kyapaz by Azerbaijan:[83] In June of 1998, Turkmenistan held its first international tender and awarded the Serdar field to Mobil. Mobil, however, declared that it could not develop the field until ownership disputes had been settled. On the Azeri side, Russian companies (LukOil and Rossneft) withdrew in July 1997 from ownership interests offered by the Azeris in its Kyapaz field because Russia supported the Turkmen claim.[84] Similar disputes between Russia and Azerbaijan will likewise slow international investment.[85] The reluctance of international investors to risk capital in cases of ownership disputes may prove to be a strong force in pushing the region toward a comprehensive demarcation settlement.[86]

Despite the substantial risks of investing in the Caspian region, foreign capital has indeed been flowing into two Caspian littoral states. Azerbaijan received a cumulative foreign direct investment (FDI) of US$2 billion from 1989 to 1997, while Kazakhstan received a larger US$4.3 billion. While these FDI totals are small relative to Poland and Hungary (at US$19 billion and US$15 billion, respectively), they are substantial compared to the other former Soviet republics.[87] These capital flows into Azerbaijan and Kazakhstan reflect the willingness of international oil concerns to come to terms with the political risks associated with authoritarian regimes.

3

Iran and Caspian Energy: Prospects for Cooperation and Conflict

Geoffrey Kemp

Introduction

For the past few years, there has been growing interest in the Caspian Basin primarily because of the belief that the region contains huge deposits of petroleum and natural gas. A related factor is that with the breakup of the Soviet Union, three of the newly independent countries that own much of the expected oil and gas – Azerbaijan, Turkmenistan and Kazakhstan – are welcoming foreign energy companies and offering them lucrative investment opportunities.

As a result, there has been talk of a "New Great Game" and a "Black Gold Rush." Over US$70 billion has been earmarked for investment in developing the energy and a major battle is being waged by neighboring countries to provide egress routes for the oil and gas to reach markets. Russia, China, the United States, Turkey, Georgia, Iran, Pakistan and Afghanistan are all putting forward proposals as to which routes for pipelines best suit the interest of the region and the world energy market. As the decisions over preferred routes reach a climax, other geopolitical factors have become embroiled in the issue, including the civil war in Afghanistan, the conflict between Armenia and Azerbaijan over Nagorno-Karabakh, secession movements and fighting in Georgia and Russia, Turkey's enduring conflict with the Kurds, and Iran's continuing confrontation with the United States. In addition to these ongoing conflicts, the region is beset with political instability, corruption and economic mismanagement. This menu of potential wealth and present problems raises a very basic question. How stable is the Caspian region and how realistic are the prospects that it will become a major player in the world energy markets? The reality is that, despite much hype and enthusiasm about the Caspian's promise, the politics of the region may ruin whatever hope there is for an open-ended bonanza. In assessing the Caspian's importance for the Gulf energy producers, it is necessary to review the very complex geopolitics of the region and the wildly different interpretations of the facts of the case.

Consider the following statement which is typical of some of the euphoric writing coming from journalists who have visited Baku and other oil regions of the Caspian in recent years:

> The region's wells will shower unimaginable wealth on people whose annual per capita gross domestic product (GDP) today hovers between US$450–US$600, building a new El Dorado in nations where camels still outnumber automobiles in 1998 . . . With many of the world's current principal oil sources drying up, the Caspian reserves are vital to the continuing prosperity of the West, and the advancement of the Western world.[1]

Contrast this statement with the words of a well-respected Central Asia specialist at the Carnegie Endowment for International Peace, Martha Brill Olcott:

> The most grandiose oil and gas projects are still in their early stages, with their viability and pace of development uncertain. The reality of post-communist development has instead been an increase in corruption and a sharp drop in living standards once protected by a comprehensive welfare net. Growing poverty has made a burgeoning trade in narcotics and illegal arms – other "global" commodities for which this part of the world is known – that are much harder to stamp out; it has also-rendered the population more susceptible to the appeal of Islamic radicalism. And ironically, many of the oil-rich region's leaders have taken the recent strategic blandishments of powers such as the United States as proof that their own position will be protected from either internal unrest or outside attack.[2]

Since there is now a growing literature on the politics of Caspian energy, this chapter will not attempt to review the entire spectrum of issues. Rather, the focus will be upon one issue that is of paramount importance to the Gulf Co-operation Council (GCC) countries and the United States, namely the role of Iran as both a Caspian and Gulf country. It will be argued that the nature of Iran's relations with the United States will have a strong influence on the debate about Caspian energy and its impact on the Gulf. First, however, it is necessary to make some general points about Caspian energy and the status of offshore ownership.

How Important is the Caspian?

The Caspian Basin is one of several *potentially* massive energy regions that *could* be developed over the next 20 years. For years energy specialists have known that eastern Siberia contains enormous oil potential but so far little has reached the international market. A combination of politics, economics and the environment has slowed development. Likewise in China's western desert, the Tarim Basin is believed to contain huge deposits, but so far the problems of egress and financing have been overwhelming. The South Atlantic, off Argentina, may be rich in oil but the dispute over the ownership of the Falklands Islands is likely to delay any efforts to explore and develop the oil potential.

In the case of the Caspian, what is known is that at least 30 billion barrels of oil are proven to be available. But this is far short of the 200 billion barrel figure often cited. The latter figure *could* be correct *once* the required drilling has taken place. But drilling requires huge rigs that have to be transported over excruciatingly difficult routes. For much of 1998, there was only one large exploratory semi-submersible drilling rig in the Caspian.[3] Furthermore, while at this time US$70 billion of investment has been *pledged* by foreign oil companies, only US$4 billion has been *spent* so far.

Three dimensions of the Caspian energy equation are of paramount concern: the ownership of the resource; the amount of hydrocarbons available for extraction; and the production and distribution costs, including direct and environmental costs. Whether there will be cooperation or conflict over access to the Caspian energy will depend upon whether these issues are resolved in an adversarial or cooperative manner.

Until 1991, the Caspian Sea was controlled by the Soviet Union and Iran. Its legal status following the breakup of the Soviet Union has yet to be defined. All five of the littoral states face major economic and political obstacles limiting their capacity to export energy. Three of the producers – Azerbaijan, Kazakhstan and Turkmenistan – are surrounded by other countries. They cannot get their energy to market without crossing another's territory. Interestingly, in the long and turbulent history of the oil business, this has rarely happened. In the past, all the great oil- and gas-exporting countries have had direct access to the world's shipping lanes.[4]

Owing to this geographic reality, the multiple routes being considered to transport the energy have assumed political and strategic, as well as economic and environmental, overtones. However, since the economic stakes for the winners are so high, the key participants are only reluctantly accepting the need to cooperate and compromise. Russia, for instance, has tried to use strong-arm tactics to assure a monopoly of the key routes from Azerbaijan and Kazakhstan. However, its ability to dictate terms has been limited because of its own financial problems and the chaos and conflict along its southern borders. Iran's problems relate to its poor relations with the United States and its difficulties attracting the investment needed to develop its huge but underutilized natural gas reserves and to exploit its geography as a transit route for its neighbors' oil and gas.

Since the choices for oil and gas routes involve so many complicated decisions, many different options have been proposed, involving countries as far apart as Bulgaria, Greece and Turkey in the west, and India, China and Pakistan in the east. Each of the many options has costs and benefits. For instance, Bulgaria and Greece, in cooperation with Russia, are proposing a route for oil transportation that would avoid Georgia, Turkey and the Bosphorus, but the plan would probably further aggravate Greek–Turkish relations. Alternatively, one of the more intriguing proposals has been to build natural gas pipelines from Turkmenistan and Iran that run into Pakistan and then on into India. Leaving aside the economics of such a project, it raises some important questions about the relationship between the subcontinent's two nuclear-armed antagonists.

The Legal Status of the Caspian Sea

The legal debate over the Caspian Sea can be traced back to the 1921 Treaty of Moscow, reaffirmed in 1935, which declared that the inland

Caspian Sea belonged to Russia and Persia.[5] Later, the Protocol of 1940 superseded the Treaty and made the Caspian joint Soviet–Iranian property.[6] Russia's position has been that the legal status of the Caspian is based on the earlier treaties between the Soviet Union and Iran and is not covered by the Law of the Sea since the Caspian is in fact a "special inner sea" and has no natural connection with any other "sea." In 1960, Iran and the Soviet Union agreed to arbitrary lines establishing territorial waters in the Caspian.

In contrast, Azerbaijan and Kazakhstan have argued that the Caspian Sea should be treated under Article 122 of the 1982 UN Convention on the Law of the Sea which defines a semi-enclosed or enclosed sea as "a gulf, basin or sea surrounded by two or more states and connected to another sea or ocean by a narrow outlet or consisting entirely or primarily of the territorial seas and exclusive economic zones of two or more coastal states." Under this definition, the Caspian would be covered by the latter qualification. Furthermore, there are precedents for dividing large lakes between two or more states (such as Lake Victoria, the Great Lakes of North America and Lake Titicaca).[7] Russia has argued that the Caspian should legally be treated as joint property shared by all the bordering states – including the newly independent republics, Russia and Iran – which would effectively delegitimize efforts by any one state to unilaterally exploit the Caspian's resources within its own demarcation lines. These lines would be drawn from the land boundries to an equidistant line in the center of the Caspian, to effectively divide up the entire sea.

Recently, however, Russia has adapted a more conciliatory tone and has sided more with Turkmenistan, Kazakhstan and Azerbaijan regarding the demarcation of the Caspian Sea. This policy change has left Iran more isolated. While Moscow has already signed a non-binding protocol with Azerbaijan and is expected to conclude a treaty with Kazakhstan soon, Iran's argument remains that it will accept unilateral exploitation of the Caspian Sea's oil and gas reserves only if the sea is sectioned into equal parts.

Russia earlier supported Iran's position on the demarcation issue. Tehran maintains that the Caspian Sea should be classified as a lake and that each of the five littoral states should have the right to develop whatever projects they choose. Iran has a strong interest in claiming a share of the seabed beyond its own sector because the main oil fields lie in the middle of the sea, off Azerbaijan. The least promising waters are those off Iran. Until there is full agreement between the littoral states on the demarcation issue, it can continue to cause delay and frustrate efforts to fully explore the potential of the offshore region.

Disputes over the Caspian are paralleled by an age-old dispute between Turkey and Russia over the Turkish straits, which continue to comprise a critical export route for energy resources from the former Soviet republics. The 1936 Montreux Treaty bars Turkey from restricting Russian merchant shipping through the Bosphorus. However, it appears that Turkey is prepared to mobilize environmentalists and world opinion to limit Russian oil transit on the waterway and will promote its own alternatives for bringing Caspian oil to the Mediterranean via pipelines that bypass the Black Sea. There is uncertainty over how far Turkey will actually go to limit tanker traffic. According to some experts, the fact that Turkey has issued a tender for a Vessel Tracking System to ensure safer traffic suggests flexibility in its position.[8] Experts point out that in 1996, 2 mt of oil flowed eastward through the Bosphorus from the Mediterranean to Bulgaria and Romania. This figure could drop considerably when Caspian oil comes on line and can be shipped to these countries without transiting the Straits.[9] There is no doubt that Turkey's environmental concerns have merit, as anyone who has crossed the Bosphorus can attest. However, the environmental issue is only one factor in the complicated competition for access routes.

The Importance of Iran

While Russia is clearly the most important Caspian player, from the perspective of the United States and the GCC countries, Iran is an equally significant player with a powerful impact on both Gulf and Caspian geopolitics and economics. How the United States and Iran resolve (or do not resolve) their problems is a significant factor influencing the future direction of Caspian and Gulf stability. A rapprochement between the US and Iran would permit economic, rather than political, criteria to determine pipeline routes and investment opportunities – unless, of course, it resulted in a more antagonistic Russia playing a tougher game. A marked deterioration in US–Iran relations could have a particularly negative impact on regional development in the Caspian and would likely increase the prospects for violence and instability. It is useful, therefore, to review Iran's relations with its Caspian neighbors and the United States.

Iran and Russia

Iran and Russia are the two most powerful Caspian countries. They also share many common interests in other parts of the greater Middle East. They both regard long-term instability in Central Asia and the Caucasus as inimical to their security and have worked carefully to make sure that other regional countries remain either stable or within their respective spheres of influence. When instability has occurred, as in Chechnya and Georgia, they have gone to great lengths not to antagonize each other. Both countries stand to benefit through projects to develop the potential energy wealth of the Caspian Basin. They still have outstanding differences on how to demarcate the offshore resources of the Caspian Basin with Iran's position more at odds with the smaller countries than Russia.[10] However, both disapprove of the efforts being made by the United States to determine the selection of pipeline routes, especially the proposed trans-Caspian routes that bypass both Russia and Iran. They share a common perception that the United States is trying to keep them out of a region of which they are a part.

It is likely that this alliance of interests will continue in the absence of a major change in US–Iran relations. It is not like the period of the Cold War when the Iranians felt threatened by Moscow and therefore sided with the West. Iranian policy toward countries of the former Soviet Union has been pragmatic and devoid of the ideology associated with its activities in other regions, including the Gulf, Lebanon and Africa. Nevertheless, over time there are potential conflicts of interest between these two major powers. Since both Iran and Russia are natural egress routes for Caspian Basin oil and gas, if the Iranians and the United States were to repair their relationship and Iran was permitted to become a key egress route, it would run into direct competition with Russia. This will be especially intense if production levels of the Caspian oil and gas fields fall short of some of the more exaggerated expectations. Even with its current system of pipelines, Iran claims that it could shift up to 700,000 b/d in the form of swaps from the Caspian countries through its own pipeline network, exporting its own oil from the Arab Gulf while using Caspian oil for its own refineries and its market in the north. Although this figure is probably exaggerated, there is no doubt that oil swaps are likely to become a significant factor in Caspian exports and that Iran will benefit from this arrangement.

Similarly, a major buildup in Iranian military capabilities, particularly if it involves long-range missiles and weapons of mass destruction, could eventually pose a threat to Russia. The testing of the Shehab-3 missile demonstrates that Iran could eventually possess the capability to reach parts of Russia. Hence, the question arises, why is Russia helping Iran in its military acquisitions? The answer is to be found in the confused and conflicting state of affairs in Moscow. As on other foreign policy issues, including relations with China, Russian policies seem to contradict each other. The Foreign Ministry and the Ministry of Defense are often at odds with aggressive lobbies pushing for arms sales and technology transfers. Russia's huge oil and gas companies (Lukoil and Gazprom) have great clout in Moscow and in many ways operate their own foreign policies.

Although Russia's Prime Minister Yevgeny Primakov has attempted to impose some discipline and a coordinated strategy on the activities of the various Russian interest groups that deal with Iran, it is not clear how successful he has been. This issue comes up frequently in the context of individual Russians and small Russian companies that have been helping Iran to develop its missile program. On each occasion when approached by American officials, the Russians deny that there is any formal government policy in favor of such help; this would be a violation of the Missile Technology Control Regime (MTCR) which Russia has signed. Yet, Russia seems to be incapable or unwilling to effectively enforce its own laws, which either reflects the weakness of Moscow or seems to be a malevolent attempt to undermine American interests by building up Iranian capabilities.

Iran and Turkey

Iran and Turkey have an important, but ambivalent relationship. There have been terrorist attacks against Turkey's secular intellectuals, journalists and politicians. During the period June 1996–June 1997, when Necmettin Erbakan held the Turkish premiership, relations between the two countries improved in view of Erbakan's Islamist credentials and his bitter criticisms of the West and Israel. However, since his ouster, political relations have become more tense, especially in view of the security cooperation between Turkey and Israel. The relationship reached the freezing point in February 1997 when Ankara declared then Iranian ambassador to Turkey Muhammad Reza as *persona non grata* after he attended a radical religious rally and allegedly criticized Turkey's secular system. Moreover, Turkey continues to fight Kurdish rebels in the southeast and has repeatedly accused Iran of supporting its outlawed Kurdish Workers Party (PKK). During the Turko-Syrian crisis of autumn 1998, also over alleged support for the PKK, Tehran repeatedly pledged its support for Damascus during the Turkish military buildup opposite the sparsely defended stretches of northern Syria.

Nevertheless, the key to Iranian–Turkish relations may be economic. With a growing population of 62 million, Turkey expects to face serious energy shortfalls in the near future. Estimates show that energy demand will rise by 200 to 300 percent in the next 10 to 15 years, and Iran may become an important supplier.[11] The proposed gas pipeline through Iran to Turkey that would bring first Turkmeni and then possibly Iranian gas to Turkey's huge market may indeed become a reality in view of the economics and Turkey's reluctance to become overly dependent on Russia for energy supplies.

Turkey's cooperation with Iran on energy projects runs counter to preferred US policy which is to support an "east–west" option to bring both oil and gas from Azerbaijan, Kazakhstan and Turkmenistan through Georgia to Turkey. This route would achieve the Clinton administration's goal of limiting major egress routes through Russia and Iran, an issue which is discussed in more detail later in this chapter.

Iran and the South Caucasus

Iran's attitude toward its neighbors in the southern Caucasus is complicated. Georgian President Eduard Shevardnadze's regime remains threatened with violent ongoing clashes in the north. Iran's relations with the former Soviet republic have traditionally been fairly cool, but some steps towards reconciliation between the two countries were made in 1998. Stressing that cooperation would strengthen security and tranquillity in the region, President Khatami called for an expansion of relations in a meeting with the Georgian Foreign Minister Iraklii Menagarichvili.[12] Menagarichvili's visit to Tehran resulted in a memorandum of understanding, stressing the importance of setting up a joint economic commission between the two countries. The two sides also emphasized the need for peaceful resolution of regional disputes, including Abkhazia, and the importance of safeguarding security in the Caucasus.[13]

Tehran has not approached Georgia's western neighbor, Azerbaijan, in the same way. Despite its potential oil riches, Azerbaijan faces deteriorating economic and social conditions. The conflict with Armenia over Nagorno-Karabakh, whose population is 75 percent ethnically Armenian, remains unresolved, and Iran continues to provide economic support to Armenia.

From Tehran's point of view, Baku's close relations with the United States and its stance on the demarcation of the Caspian Sea remain obstacles to warmer relations between the two neighbors. On the issue of pipelines, Azerbaijan's president, Gaidar Aliyev, is in favor of the Baku–Ceyhan route advocated by the Clinton administration. The Azerbaijani government has accused Iran of financially supporting a coalition of opposition parties in Azerbaijan as well as the banned Islamic Party of Azerbaijan (IPA), and four IPA leaders have been imprisoned on charges of spying for Iran.

Iran and Central Asia

Iran is struggling with its eastern neighbor, Afghanistan. Tehran opposes the Taliban regime, which has attempted, hitherto with little success, to establish complete control over the entire country while it continues to impose its ultra-fundamentalist interpretation of Islam over the part of the country it governs. The northern enclave, small but well armed (mainly by Russia), eluded the Taliban's grip for some time but lost a major stronghold in the fall of Mazar-i-Sharif in the summer of 1998. The regime has no bureaucracy, no educated cadre and not a single trained economist, engineer or oil expert;[14] at the same time, it faces physical, economic and social devastation from years of civil war.

Following the fall of Mazar-i-Sharif, it came to light that the Taliban had killed a great many Afghani Shias, people who traditionally looked to Iran for support and protection, and had kidnapped more than 40 Iranian nationals. The Taliban failed to respond to demands to turn over the Iranians, some of whom had been murdered, and Iran proceeded with a buildup of some 200,000 troops on its border with Afghanistan, under the guise of military exercises. War was averted when the Taliban released the Iranians, but only after bombastic threats from Iran and responses from the Taliban that included threats to use its scud SSMs against Iran.

Iran has established a warm relationship with Afghanistan's common neighbor, Turkmenistan. This gas-rich, Caspian state under the authoritarian leadership of President Saparmurad Niyazov is also facing an uncertain future. Economic deprivation of the vast majority of the population, worsened by burgeoning corruption, could cause social unrest if the oil dividends fail to benefit the population as a whole. Niyazov's unwillingness to enact any market reforms has translated into continued negative growth and high inflation. Under these conditions, obtaining income from exports of its natural resource is absolutely essential.

Iran has helped relieve some of Turkmenistan's economic troubles. The Korpeje–Kurt Kui gas pipeline from Turkmenistan to Iran, put into operation in December 1997, has now become the only export pipeline of Turkmenistan not dependent upon Russia.[15] In addition, a visit in July 1998 by Niyazov to Tehran resulted in the signing of a joint communiqué and three agreements on cooperation in the fields of politics, economics and sports.[16] In the communiqué, Iran and Turkmenistan voiced objection to the agreement between Russia and Kazakhstan dividing the northern sector of the Caspian Sea.[17]

The demarcation issue continues to complicate Iran's relationship with Kazakhstan, the largest Central Asian country. Like many of its CIS neighbors, Kazakhstan, led by President Nursultan Nazarbayev, is battling deteriorating economic and social conditions. The president has heightened popular resentment by spending billions of dollars on a new capital at Akmola, now renamed Astana.[18] The country's international relations are complicated by its need to perform a balancing act between China and Russia while at the same time being vulnerable to Russian pressure because of its substantial Russian minority (37 percent).

Despite US pressure, Kazakhstan's leadership has indicated that it wants to develop relations with Iran. In November 1997, the Kazakh ambassador to the United States, Bolat Nurgaliyev, said, "We believe that trade and economic contacts with Iran are very important."[19] During a visit to Tehran in March 1998, Kazakh Foreign Minster Kasymzhomart Tokayev voiced the Central Asian republic's readiness for cooperation in the oil sector.[20] In November of 1998, the Kazakh Energy Ministry reiterated its approval of oil swaps with Iran.

Iran's Energy Dilemma

Iran's relations with its Caspian neighbors are strongly influenced by its complicated energy situation, its unique geography and its continuing conflict with the United States. In order to develop its economy to meet the growing expectations of a large, young and increasingly dissatisfied population, Iran must undertake a massive capital investment in its energy sector and work closely with both its energy-rich and energy-deficient neighbors to achieve a mutually beneficial economic relationship. Mismanagement of the Iranian economy by the regime has compounded the problems to the point where even if US–Iran relations improved, US sanctions were removed and objections to pipeline routes were laid aside, it is questionable whether foreign companies will invest significant sums of money in Iran until its government enacts far-reaching reforms designed to protect the interests of foreign capital. This means opening up its oil and gas sectors to foreign investment, a step that will be regarded as anathema by the more die-hard conservatives opposed to foreigners and their money – an integral component of the revolution in its early days. Although Iran has already taken steps in this direction – most obviously in the case of the aborted Conoco deal and the follow-up agreement with Total, Gazprom and Petronas – so far the concessions have been restricted to offshore energy assets. Undoubtedly, a way will be found to get around Iranian law (which forbids foreign investment in onshore energy), primarily by negotiating "buy backs." But until the investment climate has been tested and foreign companies are assured that they will be treated fairly, the likelihood of massive investments seems slim, despite increasingly friendly overtures from the Iranian government.[21]

An analysis of Iran's energy situation helps to explain why this issue commands such a priority in Iranian politics. Iran remains one of the world's largest oil producers, but revenues have fallen significantly in recent years as production has fallen, domestic consumption has increased and the base price of oil on the world market has dropped.

In 1978, the last full year of the Shah's rule, Iran produced 4,252,000 b/d. Of this, domestic consumption accounted for 517,000 b/d (12 percent of total production), leaving 3,735,000 b/d (88 percent) for export. In 1997, production was 3,730,000 b/d of which 1,255,000 b/d (33 percent) was for domestic consumption, leaving 2,475,000 b/d (67 percent) for export. Thus, not only has production fallen but the revenues from exports as a percentage of total production has fallen from 88 percent to 67 percent.[22]

Following serious production and maintenance problems, Iran is now working to rebuild its deteriorating oil fields.[23] From 1993 through the end of 1996, Iran boosted production to an average of 3.6–3.7 mbpd, including pumping as much as 4 mbd to demonstrate productive capacity.[24] Some have suggested that such high levels of pumping will damage Iranian oil reserves. Indeed, recent reports suggest Iran has been unable to meet its OPEC quota. At these high levels, Iran's proven oil reserves will last just over 40 years. The depletion or damaging of Iran's existing reserves makes expansion of Iran's proven reserves essential. As part of its effort to get a larger share of the resource pie, Iran has attempted to be a party to deals in the Caspian Sea.

Iran's other great asset is natural gas, which is regarded as one of the most desirable sources of energy because of its abundance, competitive production costs and environmental cleanliness. The Arab Gulf countries have huge, proven natural gas reserves, especially Iran. In fact, Iran has the world's second largest proven reserves after Russia. Yet from the perspective of the Gulf countries, significant revenues from natural gas exports are very much part of their future rather than their present economic calculations. Although Iran contains the world's second largest natural gas reserves (after Russia), with nearly 16 percent of the world's total, its exports of gas are minuscule, with only 0.1 bcm sold to Azerbaijan in 1997. The remainder is used for domestic consumption and re-injection into aging oilwells to raise pressures necessary to bring the oil to the surface.

Iran's energy dilemma is compounded by its demographic distribution which determines domestic demand and the actual location of its oil and gas deposits. The major share of Iran's oil and gas wealth is located in the south and southwest of the country, whereas the majority of the Iranian population (of 65 million) is located in the big cities to the north.

It is this asymmetry between the location of the energy and the markets that makes it an attractive alternative to bring oil and gas from Iran's northern neighbors to meet Iran's northern demand and use Iran's gas in the south for distribution elsewhere, most notably to the Asian markets and, in the case of gas, to India and Pakistan.

One of the most promising developments from the Iranian point of view is the notion of swapping oil from Kazakhstan, Turkmenistan and possibly Azerbaijan and Uzbekistan for use in the Iranian market to be offset by equivalent amounts of Iranian oil lifted from its southern fields and shipped to markets in Europe and Asia. While there are complicated financial and technical matters to be worked out, including tariff costs and the compatibility of crude oil from the Caspian field with the Iranian refineries, Iran's demand for petroleum is sufficiently great for it to absorb large quantities of oil from its northern neighbors provided that this can be delivered and processed at reasonable cost. Thus, the Iranians have proposed building a new pipeline from the Caspian port of Neka to Tehran to link up with the Iranian pipeline network to permit Caspian oil to proceed to refineries at Tehran, Tabriz, Arak and various other locations in the north. This petroleum would then be refined into gasoline, kerosene and various other products that the large Iranian market could easily absorb. This would mean that oil that presently has to come from the south to the north in order to meet Iran's domestic needs can now be exported to foreign markets, and the receipts from these additional exports would then be repaid to the northern producers. Another proposal would be to bring oil from Baku to Tabriz for use in the Iranian market or if politics permit, become part of a longer pipeline extending into Turkey.

Like all issues relating to Caspian trade, a lot depends on the logistical feasibility of such movements. Oil can, after all, be moved by many methods from the producing states to the Caspian port of Neka – barge, ship, pipeline or train. Given the dire straits of the economies in these countries, any short-term additional revenue will be extremely important. This is where Iran's advantages come into play. It will take years for either the trans-Caspian pipeline or the northwestern route to Novorossisk or the eastern route to Asia to be developed. If, in the meantime, the Iranians can begin to swap oil with their neighbors, it could promote investment in Iran's infrastructure for transportation that would eventually raise the number of barrels swapped to between 400–500 thousand b/d. (Some suggest the figure could go as high as 700,000 b/d.) This is not a negligible amount of oil and could compete with oil that would otherwise traverse different routes, including the trans-Caspian, Russian or Chinese routes. Furthermore, the Iranians argue, the costs of this proposal are lower than the costs of the proposed east–west pipeline. However, it would add to the traffic flowing through the already crowded Arab Gulf and Strait of Hormuz, raising for some countries the strategic dangers of dependency on this egress route.

Current proposals to market gas out of Iran and Turkmenistan have strong geopolitical overtones. The first route concerns the shipment of Turkmeni gas to the huge market in Turkey. This would bring gas through a pipeline already completed from Turkmenistan to Kord Koy in Iran that would then link up with the existing Iranian gas pipeline system that flows westward to Neka. After the new pipeline has been completed, it would link up with the existing pipeline to Tabriz and then on through a new pipeline to Erzurum in Turkey. In the short run, this route makes a lot of sense. Iran would merely be the transit state since it would be Turkmeni gas that would be supplied to Turkey. However, over the long run, this could compete with Iranian plans to develop gas fields in the south and link them to the pipeline systems in the north. There may be a situation where over a period of years Turkmenistan and Iran could become competitors for supplying gas to Turkey. It is not yet clear what the status of the Turkmeni–Turkey pipeline via Iran will be and whether or not it will be covered by US sanctions legislation.

The other Iranian proposed pipeline that has received a lot of attention would run from the South Pars gas fields, along the coast of Iran into Pakistan and then possibly into India. This pipeline would be designed to meet the intensive gas needs of both Pakistan and India and would benefit from the relatively close proximity of the southern Iranian gas fields to the Pakistani market. However, the current political situation in India and Pakistan makes it highly unlikely that it will receive financing any time soon, but from an economic standpoint, this proposal also makes a great deal of sense.

The third option is the Turkmenistan proposal to transport gas through Afghanistan to India and Pakistan from the gas fields in Yashlar and Daulatabad. This route would be promising were it not for the fact that Afghanistan is a cauldron of instability. These three examples highlight the dilemma that is faced by all countries in the region. There is a growing regional demand for gas, particularly from Turkey, India, Pakistan and Iran, yet all three routes have serious political and financial problems that need to be addressed.[25]

US Policy on Pipelines

Against this backdrop of promise stands the United States whose current policy is to deny Iran any egress route from the Caspian across its territory. Despite its proclaimed policy of promoting multiple pipeline routes, the Clinton administration is committed to "avoiding transit routes through Iran."

The United States has a major interest in Caspian oil and gas for a number of geopolitical and economic reasons, including its growing dependency on oil imports over the next decade. Thus, like all industrial powers, the United States wants to guard against energy supply disruptions and seeks international cooperation on global energy issues.[26] It follows from the overall objectives that the core of the administration's energy policy in the Caspian region is the multiple pipeline strategy. According to Energy Secretary Federico Peña, multiple pipeline routes "will enhance the commercial interests of companies operating in the Caspian states as well as host states."

However, the United States does not want Russia to unilaterally decide on the rules of the Caspian oil and gas game, and it also intends to exclude Iran from the game. Thus, the United States is pressing for an "east–west energy corridor," i.e. a Baku–Ceyhan line with a trans-Caspian extension to Turkmenistan and Kazakhstan.

In a statement to the House Committee on International Relations, April 30, 1998, Secretary Peña succinctly reviewed the US policy in the Caspian region:

> The Administration is committed to an "east–west energy corridor" to link the countries in the region to the west. Our east–west energy corridor initiative principally supports the construction of trans-Caspian oil and gas pipelines running under the Caspian Sea, linking the countries on the eastern Caspian shore – Kazakhstan and Turkmenistan – with those on the west, starting with Azerbaijan in the coastal city of Baku. Both lines would traverse Georgia. The oil line would proceed to the Mediterranean port of Ceyhan, while the gas line would serve Turkey's growing gas market.[27]

The proposed US trans-Caspian corridor would bring both oil and gas across the Caspian from Central Asia to Azerbaijan and then onto markets – gas in Turkey and oil to Ceyhan on the Mediterranean. The precise routes for crossing the Caspian have not yet been worked out – a feasibility study is currently underway. There remain overwhelming political and environmental problems, and the economic viability of the trans-Caspian is doubted by many. In order for the project to be worthwhile, it would only make sense if it included both oil and gas and if it took the majority of oil and gas produced by Kazakhstan, Uzbekistan and Turkmenistan across into Azerbaijan. This would most likely work out if the northern route to Russia and Novorossisk or the southern routes through Iran or Afghanistan do not work out. Thus, the irony is that in developing a plan for a trans-Caspian pipeline, the United States is essentially ruling out other multiple egress routes, which is counter to its original strategy. Furthermore, if the pipelines were built, it would endow Azerbaijan with considerable strategic importance since it would be the junction for not only its own oil but also oil and gas coming from its neighbors across the Caspian.

There are other factors that suggest that the administration ought to reconsider its policy, both in terms of the overall energy equation in the Caspian Basin and in terms of its relationship with Iran. First, recent reports have questioned the potential of the region. A Baker Institute study said that even under the most optimistic assessments, by 2010 Caspian oil production will likely reach little more than 3.5 mb/d and cover only 3 to 4 percent of expected worldwide use.[28] This is due to a range of geological, logistical and political challenges, some of which were discussed above. The study concludes that there should be only one main export line; in other words, the multiple pipeline strategy is unsustainable. The International Institute for Strategic Studies (IISS) in London followed up by stating that the Department of Energy's projection of around 200 billion barrels of oil in the Caspian Sea was "widely derided" in the oil industry. It said most forecasts showed the region contained between 25 billion and 35 billion barrels.[29] The low oil prices make the oil companies even less eager to invest in multi-billion dollar pipeline projects. In short, multiple pipeline routes may *not* enhance the "commercial interests of companies operating in the Caspian region."

Second, given rapidly deteriorating social and economic conditions in the region, the littoral states are eager to obtain their much-needed oil income.[30] As noted above, for Kazakhstan and Turkmenistan, the quickest and cheapest way is to go through Iran. Turkmenistan is especially eager to export its gas and oil to the Persian state. Turkmenistan's deputy foreign minister for economic affairs, Yolbaz Kepbanov, said, "Let the Americans say, 'Don't be friends with Iran,' but we can't do that, because we are a neutral country. In our eyes, everyone is equal but Iran comes first . . . We have to cooperate with them."[31]

Third, the arguments for an Iran option are not only logistical, financial and ecological, but in some ways also geopolitical. In 1998, Iran defaulted on repayments of its debt to international creditors for the first time in four years and is currently suffering the results of low oil prices and economic mismanagement. Nevertheless, the Iranian regime is more stable than most of the Caspian newly independent states, enjoys a degree of domestic legitimacy and, the American embargo not withstanding, has seen a general improvement in its international position. A European official summed up the situation by saying, "The US position of excluding Iran is patently untenable."[32]

In the final analysis, the impact of Caspian energy on the security and prosperity of the Gulf states is likely to be political rather than economic. It is true that greater oil production from the Caspian will impact oil prices, but to what extent, will depend upon prevailing market conditions. It should also be noted that an end to sanctions against Iraq and its rehabilitation as an oil producer could have a very significant impact on prices. Furthermore, some analysts believe that there is much new oil to be discovered and extracted from Iraq. If oil prices – and therefore profit margins – remain low for the next few years, Iraq rather than the Caspian could become the focus of a "Black Gold Rush." The precarious state of the political economies of the Caspian countries suggests that this scenario is one to be feared, especially if a new regime in Iraq emerges that is predisposed toward the West and the free market.

4

Political Trends in the New Caspian States: Azerbaijan, Kazakhstan and Turkmenistan

Shirin Akiner

Introduction

In the early nineteenth century Russia, having annexed most of trans-Caucasia and much of western Kazakhstan, became the dominant power in the Caspian Sea. The Gulistan Pact of 1813 and the Treaty of Turkmanchai of 1828 allowed Iran some rights of access to the sea, but gave Russia the exclusive right to maintain a fleet of warships. After the Bolshevik Revolution, these unequal arrangements (always much resented in Iran) were abrogated, to be replaced in 1921 by the Iran–Soviet Friendship Treaty. In theory, this gave equal rights of access to the two signatories, but in practice Russia, and subsequently the Soviet Union, continued to play the dominant role. In trans-Caucasia and Central Asia, there were brief struggles for national independence, but by the early 1920s, Soviet rule had been established in these regions. Hence, from this period until the collapse of the Soviet Union at the end of 1991, there were only two state entities bordering the Caspian Sea – the Soviet Union and Iran.

With the disintegration of the Soviet Union, three new littoral states came into existence: Azerbaijan, Kazakhstan and Turkmenistan. Today, these states are grappling with the complexities of rapid economic and social transition. They are also faced with the need to construct a new political order. The responses of these new states to the challenges of the post-Soviet era have many common features. This is not surprising, since they have had similar experiences over the past seventy-odd years of Soviet rule. Moreover, the titular peoples in each of these states are ethnically Turkic and by religion (at least nominally) Muslim. Yet increasingly, there is a divergence in the pace and style of reform. This is partly rooted in different historical experiences. Partly, too, it is the result of geographic location, which in turn determines the opportunities for contact with the global community, and hence exposure to external influences and pressures. There are also significant differences in land area and population, likewise in social and ethnic composition. All these elements are currently interacting in dynamic fusion, and influencing – often erratically – the process of change.

The problems of transition are not, of course, unique to this region. However, here they attract particular attention because these countries, as well as the Caspian Sea itself, possess major reserves of hydrocarbons. Foreign oil and gas companies have already made large-scale investments in research and development. Competing export pipeline projects are the focus of fierce lobbying at the national and international level. Whether it will be possible to continue to attract the foreign investment that is required in order to exploit fully the economic potential of this region will depend on the level of political risk. This in turn will largely be determined by internal developments in these new countries. This chapter examines the principal features of the emergent post-Soviet political systems in Azerbaijan, Kazakhstan and Turkmenistan and attempts to draw some conclusions about likely future trends. (For background information see Appendices 1 and 2).

Independence Impact

In order to set post-Soviet developments in Azerbaijan, Kazakhstan and Turkmenistan in context, it is necessary to bear in mind the situation at the time of the collapse of the Soviet Union. In August 1991, conservative forces in Moscow staged a *coup d'état* to remove President Gorbachev from power. The attempt failed, but it accelerated the process of disintegration that culminated in the formal dissolution of the Soviet Union a few months later. Several Soviet republics declared their independence during this period, including Azerbaijan (August 30)[1] and Turkmenistan (October 27); Kazakhstan was the last to follow suit (December 16), but only after it became clear that the Soviet Union had effectively ceased to exist.[2]

In Turkmenistan and Kazakhstan prior to the events of August 1991, there had been no liberation movements or demands for autonomy. On the contrary, there was considerable support for remaining within the Soviet Union.[3] Consequently, these countries were in no way prepared for independence when it came. They were still bound to the "center" psychologically, as well as by a myriad formal and informal links. The suddenness with which the Union collapsed could well have led to turmoil, but in fact, the constitutional transition from Soviet republic to independent state took place almost imperceptibly. This was largely because the leadership in these republics remained unaltered and firmly in control. In the political sphere, there was thus continuity and this helped to mitigate the sense of shock and uncertainty.

The situation in Azerbaijan was very different. Here a sense of national awareness, and concomitantly of anti-Soviet sentiment, had crystallized before the collapse of the Soviet Union. It was prompted by the war over Karabakh. The sequence of events that led to the outbreak of hostilities began in 1987, when the Armenian population of Karabakh, an autonomous province within Azerbaijan, actively began to demand unification with Armenia. This provoked a furious reaction from the Azeris and in 1989 led to a massacre of Armenians in Azerbaijan (nearly half a million Armenians were domiciled in Azerbaijan at this time). The Armenian government retaliated by expelling thousands of Azeris from its territory. There were further attacks on the Armenian population in Baku, the Azeri capital, in early 1990. The central government sent in troops to restore law and order and in the ensuing clashes many Azeris were killed. On January 6, 1992, Karabakh declared its independence. A full-scale war broke out between Azerbaijan and the Karabakh Armenians. The latter made rapid gains. The newly formed Azeri army achieved some successes, but by early 1993 the Karabakh Armenians were once more in the ascendancy and before the end of the year they had taken control of the entire southwestern part of Azerbaijan. By May 1994, an estimated 25,000 people had been killed in the war. A cease-fire was declared and international peacekeeping initiatives were launched. At the time of writing, the cease-fire remains in place, but there has been no final resolution of the conflict.[4]

Anti-Armenian feeling, as well as anger over Russia's perceived support for the Karabakh Armenians, precipitated a wave of nationalism in Azerbaijan. The Popular Front movement emerged as a powerful political force in the late 1980s. Fervently anti-Communist in outlook, it began to re-assert ideas of Azeri identity and to lead the struggle against Soviet domination. Other nationalist groups were also formed at this time, some consciously drawing on the legacy of the brief era of pre-Soviet national independence. However, Communist Party officials continued to dominate the government. It was only after the defeat of Azeri troops in 1992 that the leader of the Popular Front, Abulfaz Elchibey, was able to make a bid for power. He was elected president in June that year on an anti-Russian, pro-Turkish platform. He promised democratic reforms within the country, and victory in Karabakh. He did not succeed in delivering either, and the country soon began to slip into chaos. By 1993, he had lost control of the situation and was forced from office, to be succeeded by a former Communist leader, Haidar Aliyev. Aliyev restored law and order domestically and in his foreign policy pursued a more even-handed approach in relations with neighboring countries. A cordial attitude was adopted toward Russia and Iran, though strategic distance continued to be retained; relations with Turkey remained good, but were given less prominence than under the previous leader.

The economic consequences of the disintegration of the Soviet Union were severe in all three countries. Previously, they had formed part of a single economic space, a single economic system. As other Soviet republics, they did not have separate national economies. When the Soviet Union collapsed, budgetary transfers from the central government suddenly ceased, causing serious shortfalls in the funding of social welfare services. Trade and transport links were also abruptly dislocated, leading to massive bottlenecks and shortages. Hasty and ill-prepared attempts at economic reform exacerbated the problems. Soon there was hyperinflation and soaring unemployment. Manufacturing industries ground to a halt and the agricultural sector was devastated. Poverty and deprivation became widespread. In Azerbaijan, the cost of the war with Karabakh and the influx of nearly a million refugees created crippling additional burdens.[5]

However, there was general optimism about the future, as it was believed, at the government as well as the popular level, that exploitation of the region's rich hydrocarbon resources would provide a solution to its economic woes. Since independence, major oil and gas deals have been concluded with international consortiums, but to date, revenues have fallen far short of local expectations. Reasons for this include factors such as problems over export routes; lack of clarity over the legal status of the Caspian Sea, hampering offshore developments; and low world prices for hydrocarbons.[6] These difficulties will no doubt be overcome eventually, hence the longer-term economic outlook for these countries remains favorable. In the shorter term, however, the rate of recovery is likely to be slow. Consequently, there will probably not be a significant improvement in the living standards of the population at large in the near future.

Leadership Profiles

Since the collapse of the Soviet Union, the dominant feature of the political systems in Azerbaijan, Kazakhstan and Turkmenistan has undoubtedly been the role of the leader. Power today is overwhelmingly concentrated in the hands of the incumbent presidents – Haidar Aliyev, Nursultan Nazarbayev and Saparmurad Niyazov respectively. All three came to prominence during the Soviet period. They were groomed within the Communist Party hierarchy and acquired legitimacy and authority as they rose through the ranks. Equally important, they honed their survival skills in the harsh school of intrigue and competing interests that characterized the system. Their career paths during this period are superficially similar, in that they held important posts both within their own republics as well as in Moscow, but the focus of their activities was in somewhat different spheres.

Haidar Aliyev (b. 1923) spent his early career in the KGB, eventually becoming the first Azeri to head the KGB of Azerbaijan (a post previously reserved for Slavs). A Brezhnev protégé, he served as First Secretary of the Communist Party of Azerbaijan from 1969 until 1982 and thereafter, under Yuri Andropov, was elevated to the central party leadership as a full member of the Politburo. In 1987, under Gorbachev, he fell from favor and was removed from this post. His political prospects looked bleak. He subsequently returned to Azerbaijan and for a time headed the local parliament in his home province of Nakhichevan. During this period however, far from withdrawing from the national scene, he consolidated his power base within the republic. The turmoil that ensued following the collapse of the Soviet Union and the Azeri military fiasco in Karabakh provided him with the opportunity to return to power. In 1993, he was elected president of Azerbaijan with a vast majority.[7] He was re-elected in 1998 with a smaller, but still rather substantial share of the vote.[8]

Nursultan Nazarbayev (b. 1940) worked in industry for several years before turning to full-time employment in the Communist Party administration. In 1984, he became Chairman of the Council of Ministers of Kazakhstan. His real rise to prominence, however, came under Mikhail Gorbachev. In 1986, he was made a member of the Central Committee of the Communist Party of the Soviet Union, based in Moscow. Concurrently, from 1989 to 1991, he served as First Secretary of the Communist Party of Kazakhstan. In 1990, he became the first President of Kazakhstan (still part of the Soviet Union), elected unopposed to this post by the Supreme Soviet of the republic. In early 1991, it was widely thought that he might be elevated to some prominent Soviet role, possibly as deputy to President Gorbachev. Before the end of the year, however, the Soviet Union had ceased to exist. In December 1991, he was reconfirmed as president of Kazakhstan by a nationwide (though uncontested) general election for five years. In 1995, though, his term of office was prolonged by referendum to the year 2000. Snap elections were held in January 1999 and he was re-elected for a term of office that was now extended to seven years.[9]

Saparmurad Niyazov (b. 1940) spent most of his early career working in the Communist Party apparatus in Leningrad (where he had studied) and in Turkmenistan. From the mid-1980s, he concentrated on accumulating power in Turkmenistan. Like Nazarbayev, he was a Gorbachev protégé. In 1985, he became Chairman of the Council of Ministers and from 1986 to 1991 was First Secretary of the Communist Party of Turkmenistan; in 1990 he also became Chairman of the Supreme Soviet of Turkmenistan and, a few months later, President. He was re-elected in an uncontested poll in 1992. In January 1994, his term of office was extended by referendum to the year 2002.[10]

After the collapse of the Soviet Union, Aliyev, Nazarbayev and Niyazov skillfully transformed themselves from Soviet officials to national heroes. They had not been freedom fighters, but rather, passive observers of a chain of events over which they had no control. Nevertheless, they succeeded in constructing images of themselves as "founding fathers" of the new states, and, moreover, as the unique guarantors of independence, stability and communal harmony. There has certainly been a notable degree of civic order and an impressive lack of friction under their rule. Detractors claim that this would have been the case whether or not these individuals had been in power. It is impossible to know, but it seems likely that the strong leadership that they have exercised has helped to maintain equilibrium.

Their record on political reform has been less positive. All have vociferously proclaimed their commitment to democratization and pluralism. However, there has to date been little real movement toward creating more liberal societies. The constitutions of these countries originally gave the president wide powers, but also set certain limitations, including constraints on age and on number of terms of office. These have been gradually removed, amended by decrees and by referendums.[11] At the same time, the incumbents rapidly established personal control over virtually every sphere of public life in their respective states. All have promoted members of their family and "clan" (friends, relations and clients) to key positions in the state apparatus, the legal system, the economy and the media. They have also stacked the bureaucracy and the provincial governments with their supporters. Thus, they have created a core network of loyal henchmen. A massive personality cult has been orchestrated around each of them, marked by numerous portraits in public places and adulatory commentaries in the official media. This trend is most marked in Turkmenistan,[12] but is also evident in the other two countries.

Opponents and critics have sometimes described the presidents of Azerbaijan, Kazakhstan and Turkmenistan as dictators. They undoubtedly resemble the absolutist heads of state that are found in many parts of the developing world rather more than they do the constitution-bound presidents of Western countries. Yet compared with the brutal rulers that have come to power in some countries in Asia and Africa, they head relatively mild regimes. They have certainly established authoritarian monocracies. Policy making – insofar as it is possible to identify how and where this process takes place – appears to be solely in the hands of the leader. Coercion is used to maintain public support for the incumbent regimes, though this generally takes an indirect form, for example, through manipulation of the media and the selective use of the law as a means of curbing incipient opposition.[13]

The security services remain, as during the Soviet period, an important tool for checking and controlling social behavior. Fear of reprisals – often an imagined threat rather than an actual danger (a legacy of the random terror of the Soviet past) – acts as a powerful incentive to concur with official policies and to accept unquestioningly the explanations provided by the authorities. Instances of violations of human rights, most frequently in connection with the suppression of political opposition, are widespread in all three countries. However, they do not extend to gross abuses such as the systematic use of torture or arbitrary executions.[14] The presidents do recognize, if only as a matter of political expediency, the need to work with international organizations in order to establish standards that are in line with commonly accepted norms. Progress is certainly slow, but in each of the three countries there has been some improvement in this area over the past few years.

Institution Building

Post-Soviet institution building has been largely driven and shaped by the incumbent leaders. There are two main strands to the process. One has focused on the consolidation of new state and national identities through the creation of symbols, rituals and histories. The other has sought to create new constitutional frameworks and organs of state management. In Turkmenistan, President Niyazov has been the guiding force in both areas. In Kazakhstan, likewise, President Nazarbayev has played an influential, though generally less high-profile, role in developing new institutions. In Azerbaijan, the process of national rebirth began before President Aliyev came to power and to some extent occurred spontaneously. In the shaping of the present constitution and state organs, however, he has had considerable influence. In Azerbaijan and Kazakhstan, Western democratic institutions have provided the theoretical model for constitutional and parliamentary reforms. The way in which those models are applied in practice, though, still owes much to the Soviet past.

Symbols and Historicization

Azerbaijan was the first of these three countries to reject the Soviet-era state symbols. In their place, the flag, emblem and national anthem of the first independent Azerbaijan republic (1918–20) were re-instated. The nationalizing project included the renaming of cities, streets, public buildings, institutions and titles, reverting wherever possible to pre-Soviet designations. An analogous process took place, though more gradually and starting slightly later, in Kazakhstan and Turkmenistan. New flags, emblems and anthems were devised, aimed at encapsulating the nation's cultural heritage.[15] There was also extensive renaming of places and institutions.[16] National currencies were adopted in 1992–93.[17] In all three countries, artists and scholars have provided intellectual validation for the new identities. In particular, there has been a proliferation of revisionist histories that re-interpret the past in a more nationalistic manner, particularly episodes demonstrating resistance to Tsarist or Soviet rule.

Language Policies

During the Soviet era, Russian, though not officially the state language, was widely used in public statements. In urban areas, it was common for people of all ethnic origins to use Russian in place of their native tongue. Toward the end of the 1980s, under the more liberal regime of Gorbachev, there was a reaction against this policy and the public use of the language of the titular people of the republic was energetically encouraged. This trend continued after independence, when it was given additional legal and administrative backing.

This nationalist trend met with the greatest success in Azerbaijan where, in the wake of resentment over Moscow's perceived support for the Armenians, there was a marked move to reject the Russian language at all levels of society. This was particularly striking in Baku, which had long been a predominantly Russian-speaking city. The influx of Azeri refugees further boosted the use of the language. In Turkmenistan and Kazakhstan, efforts to increase the use of Turkmen and Kazakh respectively have been haphazard and have made only a limited impact. There is undoubtedly greater public use of these languages today than was the case at the time of independence, but Russian continues to be widely used, particularly in urban areas.

An important symbolic step has been the decision, adopted by all three countries, to shift to the Latin script in place of the Cyrillic script, which was introduced during the Soviet period as the medium for writing Azeri, Kazakh and Turkmen. Azerbaijan has been implementing the change quite rapidly, while Turkmenistan has favored a more gradualist approach; in Kazakhstan, however, it has been quietly dropped from the current agenda.

Islam and "National Ideologies"

In the Soviet era, Marxism-Leninism, the official ideology, was ever-present, an integral feature of the very fabric of society. It shaped societal values and lent legitimacy to the regime; it explained the past, gave meaning to the present, and set goals for the future. When the Soviet Union collapsed, the ideology, too, was discredited in the eyes of all but the staunchest Communists. For society at large, this left a sudden vacuum – a loss of a sense of purpose and direction. Intellectuals and government officials soon came to regard the fashioning of post-Soviet "national ideologies" as a real and important challenge. In the absence of a clear political doctrine, there was a tendency to look to history and culture for defining features.

In all three countries, given that the titular people were Muslim, Islam was regarded as an essential component of a new "national ideology." Under Soviet rule, Islam, like all other religions, had been severely persecuted. Consequently, by the 1980s, knowledge of the faith was very rudimentary; for the great majority, it was more of a cultural identity than an active commitment. Moreover, it had been relegated chiefly to the private sphere, to the celebration of life-cycle rituals.[18] Now, as one of the symbols of independence, the state authorities are giving greater public prominence to the religion. Senior clerics are invited to take part in state events and there is substantial consultation between government and religious authorities. Many new mosques and *madrassah* have been opened. Restrictions on making the *hajj* have been relaxed. Also, it is now relatively simple to get official permission and funding to study in Muslim institutions abroad.

However, the intention is certainly not to create Islamic theocracies. First, the constitutions of these countries stress the separation of religion and state. If that position can be said to have been eroded, it is because of the level of control that the state exercises over all religious establishments and functionaries. Moreover, Muslim communities are kept under strict observation and activities such as informal prayer meetings that are not sanctioned in advance by the authorities are liable to severe penalties. Secondly, although Islam is given particular prominence, other established religions (such as Orthodox Christianity) are treated with respect and, on occasion, senior representatives of these faiths are invited to participate in public ceremonies.

Other elements that have been co-opted into the national ideologies include mythologized histories that illustrate such features as national unity and indigenous forms of democracy. Thus, for example, in Turkmenistan, much emphasis is given to Maghtumkuli, an eighteenth-century poet who called on the Turkmen tribes to abandon internecine strife and to unite; similarly in Kazakhstan, several dissertations have been devoted to tracing the roots of Kazakh democracy, delving some 2,000 years or more into the past. Ceremonies and rituals have been devised, especially in Turkmenistan, to replace the parades, rallies and other displays of public loyalty of the Soviet era.[19]

These efforts are still at an early stage, however, and the themes are still being elaborated. One problem that has emerged is how to balance the desire to create an ideology that is distinctively national, which draws on the culture and history of the titular people, but at the same time does not exclude those who do not belong to the titular ethnic group. In Azerbaijan and Turkmenistan, this is less of a problem, since the majority of the population are Azeri and Turkmen respectively. Also, in Turkmenistan in particular, positive efforts have been made to give national ceremonies an inclusive, multi-ethnic character. In Kazakhstan, however, where the Kazakhs constitute less than half the population, the situation is more complicated. The very designations used for the population of the state as a whole raise concerns: "Kazakh," for example, may be understood as referring to a specific ethnic group, while "Kazakhstani" is less ethnically specific. President Nazarbayev insists on the need to create a civic identity that embraces all the republic's citizens, but Kazakh nationalists (still a relatively small group) favor a more discriminatory policy.

Constitutions and Government Systems

Turkmenistan was the first of the three countries to adopt a new constitution and a new system of government. The new constitution, adopted in 1992 to replace the Soviet Constitution, specifies a secular democracy with a presidential system of government. There is a symbolic separation of powers, but the strong emphasis on the executive role of the president makes this little more than a formality. The legislative body (the *Majlis*) has 50 seats. The role of the deputies is primarily to confirm decisions taken by the president; there is no opposition faction.

An innovative institution created after independence is the People's Council. Its members include senior ministers, senior judges, provincial governors (all appointed by the president), deputies from each region of the country, and the president himself. The People's Council has among its formal powers the right to declare war and to amend the Constitution.[20] Another new institution is the Council of Elders. The members of this body are respected members of society, chosen from all parts of the country. The intention is that they should provide wise guidance in affairs of state, and also ensure a regional balance of interests. This Council has among its powers the exclusive right to nominate presidential candidates. The objectivity and independence of the Council may, however, be questioned, since the president personally selects all the members. The judicial system, which comprises a Supreme Court, an Economic Court, provincial courts and district courts, is likewise firmly under presidential patronage.

Constitutional and parliamentary reform in Kazakhstan was instituted slightly later. It is now officially described as a "parliamentary republic." However, over the past few years, the Kazakh parliament has seen a steady diminution of its authority, and a corresponding growth in presidential power. In 1993, the parliament that had been inherited from the Soviet era was dissolved, ostensibly as a prelude to institutional reform though more probably because the deputies had obstructed some of the president's attempts at reform. Elections to the new parliament were held in 1994. A year later, though, it was disbanded yet again, this time on the grounds of a procedural irregularity. Opposition deputies believed that this was because the legislative body was still not compliant enough.[21] The president ruled alone, by decree, for the next nine months. A new, bicameral structure for parliament was approved in October 1995; this provided for a 47-seat upper chamber (seven of which were reserved for presidential nominees) and a 67-seat lower chamber. The elections, held at the end of the year, were heavily weighted in favor of pro-president candidates. At this time, too, the president's personal staff was enlarged, while some of the ministries and state committees were pared down. Thus, within four years of independence, a style of government had been instituted that, though structurally different from that in Turkmenistan, had nevertheless created a high degree of presidential control.

Azerbaijan initially seemed set to follow a different trajectory from the other two states. This was chiefly a consequence of the turmoil caused by the war over Karabakh, which brought about two presidential elections within a year and several changes of policy. On the one hand this created uncertainty and instability, but on the other, it did allow a certain political dialogue to emerge. There appeared to be the beginnings of a genuinely pluralistic system. This trend was interrupted when Haidar Aliyev came to power in 1993. A new, more democratic constitution was approved by a nationwide referendum in 1995, but many of its more liberal articles have since been ignored. A new unicameral, 100-seat parliament was created to replace the 50-seat legislative body (*Milli mejlis*) that had been set up after independence. Elections were held later that year. However, the electoral procedures were heavily flawed and many violations were reported.[22] Almost three-quarters of the candidates who represented opposition groups were denied registration. The outcome was a parliament that was overwhelmingly pro-president in composition.

Political Institutions and Trends

Political Parties

Contemporary political life in Azerbaijan, Kazakhstan and Turkmenistan is marked by profound apathy on the part of the majority of the population. This is to some extent a reaction to the enforced politicization of the Soviet era. It is also a result of the disillusionment with the Communist Party that followed the collapse of the Soviet Union; this has engendered a sense of cynicism and distrust towards all forms of political activity. In addition, preoccupation with the problems of everyday life – for many, the sheer struggle to make a living – allows little time or energy for political issues.

Nevertheless, embryonic multi-party systems have begun to emerge in Kazakhstan and Azerbaijan. Parties tend to be centered around an individual who has a strong personality and/or sufficient wealth to establish a power base. Programs and party platforms, insofar as they are discussed at all, are generally vague, consisting of little more than idealistic platitudes. Party membership is almost always very small and restricted in its social range, drawing predominantly on a network of personal contacts and acquaintances.

Party politics do not play a significant role in either parliamentary or presidential elections: candidates at both levels are elected as individuals rather than as representatives of a particular party. In all three states, most parties have either been created by the incumbent leaders, or profess support for them. All parties and movements must be officially registered, otherwise they are declared illegal and liable to prosecution. The preconditions for registration are onerous and in Azerbaijan and Kazakhstan they are difficult to fulfill without government backing. In Turkmenistan, registration without official support is impossible.

Azerbaijan has, comparatively speaking, the most developed political system and the largest number of parties. In mid-1995, in preparation for the parliamentary elections that were to be held later that year, all political parties had to apply for re-registration. Registration was granted to 31 parties.[23] Of these, only eight qualified to take part in the elections. The party which eventually gained the overwhelming majority of seats was New Azerbaijan, founded by Heidar Aliyev in December 1992. Other pro-government parties included the Azerbaijan Democratic Independence Party, the Motherland Party, the Democratic Entrepreneurs Party, the Alliance in the Name of Azerbaijan, and the Azerbaijan National Statehood Party. The Popular Front, the party of Abulfaz Elchibey, a former president, and the Azerbaijan National Independence Party provided the main opposition. There is also a reformed Communist Party, but it plays an insignificant role.[24]

In Kazakhstan, there are currently some 10 political parties, but most are very small and almost all are pro-government. Insofar as they have programs at all, these are generally vague statements in support of health and happiness for all. The main party (since March 1999) is the Otan-Fatherland Party, which is an amalgamation of several other parties, including the Party of People's Unity of Kazakhstan. The revived Communist Party has had some success, but remains relatively small.[25]

In Turkmenistan, there have been only two registered parties: the current Democratic Party of Turkmenistan, formerly the Communist Party (renamed in November 1991), and the stillborn Peasant Justice Party. Both are sponsored by the government. In 1990, while still under Soviet rule, there was an attempt to create an independent Democratic Party. In 1991, however, the then Communist Party appropriated this designation. Consequently, the original Democratic Party changed its name to the Party of Democratic Development; it was never allowed to register and its leaders were harassed (two were reportedly incarcerated in psychiatric hospitals for a time).[26] This party is now moribund. There are a number of

Turkmen opposition groups abroad, mostly in Scandinavia, Russia and the United States, but they have very little influence within the country.[27]

Other Groups and Movements

Some signs of emergent pluralism in these countries might be seen in the recent proliferation of various types of semi-formal associations, unions and movements. Several non-governmental groups have now been registered in Azerbaijan and Kazakhstan, and a couple even in Turkmenistan. Most are concerned with issues such as human rights, gender rights and ecological problems. In general, they are closely supervised by the authorities. Their activities are hampered by the fact that they have limited and mainly short-term funding. Like the formal political parties, they attract very little public support and are frequently regarded with suspicion, as though they were of subversive intent. Nonetheless, especially in Azerbaijan and Kazakhstan, they have begun to raise public awareness and to make some contribution to policy debates.

Religious organizations in Azerbaijan and Kazakhstan, like political parties, have to be registered. The faiths that receive particular support from the state authorities are the established confessions of the main ethnic groups, namely Islam and Russian Orthodox Christianity. In recent years, Muslim and Christian missionaries of many different sects (and funded from a wide variety of sources) have made an appearance in these states; representatives of "new-age" faiths have also been active. Such newcomers are often subjected to considerable restrictions and even harassment. Unsanctioned activities, particularly among Muslims, are treated as criminal offenses because the state authorities fear the emergence of Islamic fundamentalist groups. Yet such fears appear to be exaggerated. It is difficult to know quite how many such groups might be in existence, but the level of religious observance amongst the population at large is still at a comparatively low level. In Turkmenistan, the registration of religious parties is banned; only the official religious organs (Muslim and Christian) are allowed to function.

Ethnic relations in these countries are, at least on the surface, amicable. In Azerbaijan, apart from the crisis with the Karabakh Armenians, there was a period of tension in the early 1990s and some ethnically based organizations, notably those of the Lezghi and Talysh minorities, made bids for autonomy. Since 1994, however, they have been quiescent.[28] In

Kazakhstan, there are several national cultural associations (e.g. Korean, German), but only the Russians based in the north of the country have, through their organization *Lad* ("Harmony"), sought political representation in parliament; up to about 1995 they were quite active, but over the last few years have adopted a more low-key stance. In Azerbaijan and Kazakhstan there are a handful of xenophobic, ultra-nationalistic movements among the titular peoples. They are in favor of the exodus of all other ethnic groups from their territories. They enjoy little popular support. However, in Kazakhstan they have succeeded in exerting some influence on the government and thereby gained a degree of preferential treatment for the Kazakh language and culture. The 1992 Turkmen Constitution bans political parties based on ethnic affiliation.

Military Developments

In many parts of the developing world, the armed forces play a significant role in the political process, acting as power brokers and at times taking power into their own hands. It is, however, unlikely that this will happen in Azerbaijan, Kazakhstan and Turkmenistan in the near future. The military and other armed forces in these states are still at an early stage of development.[29] Relatively small in size, they were formed on the basis of the Soviet equipment and personnel stationed on their respective territories at the time of the breakup of the Union. None of these countries possess nuclear weapons. Kazakhstan, which did have a significant part of the Soviet arsenal located on its territory, voluntarily renounced this in 1995.[30] The nuclear weapons were either returned to Russia or destroyed, in accordance with Kazakhstan's international treaty obligations.

All the new armed forces are in need of modernization, re-organization, re-equipment and re-training. There is currently recruitment of professional servicemen, as well as compulsory conscription. There are some training facilities in these countries, but also several bilateral agreements that provide for training abroad (e.g. in Turkey and some European countries). There are some high-ranking ex-Soviet officers of national ethnic origin (i.e. Azeri, Kazakh and Turkmen), but they are too few in numbers for present needs. Consequently, in Turkmenistan and Kazakhstan, some Russian officers have been retained on contract while the fledgling national armed forces are being developed. Azerbaijan has taken a determined stand on the immediate nationalization of its armed forces and no longer has any Russian officers.

The only force to have seen active service since the collapse of the Soviet Union is the Azeri army. It has not to date acquitted itself with great distinction. It has lacked effective leadership and there have been numerous allegations of corruption and internal rivalries. Numerous changes of senior personnel do not appear to have improved the situation.[31] In Kazakhstan, too, there have been allegations of corruption in the armed forces. In both countries, there are problems with draft dodging and desertion. Little is known of the situation in Turkmenistan.

In theory, all three countries support the principle of parliamentary control over the new national armed forces; parliamentary deputies in Azerbaijan and Kazakhstan, for example, have been quite active in raising questions relating to the welfare of the troops. Nevertheless, there is still much secrecy regarding policy, budget and organizational matters. Real control appears to lie in the hands of the heads of state: it is they who appoint and dismiss senior officers to positions of responsibility and thus are able to monitor and reinforce ties of personal loyalty.

Print and Broadcast Media

The Constitutions of Azerbaijan and Kazakhstan guarantee freedom of word, thought and information.[32] The Constitution of Turkmenistan does not specify such freedoms; on the contrary, it is enshrined in law that those who disseminate material that might be considered critical of official policies "commit criminal deeds" and are therefore liable to prosecution.[33] On this basis, the situation in Azerbaijan and Kazakhstan would appear to be more liberal than in Turkmenistan.

This liberal impression is further enhanced by the fact that there are far more media organizations – newspapers, magazines, television and radio companies and information agencies – in Azerbaijan and Kazakhstan than in Turkmenistan.[34] In fact, however, when actual output is examined, the differences are not so great as the legal framework would imply.

First, in both Azerbaijan and Kazakhstan, a large proportion of the communication media, both electronic and print, is either state owned or state-run. Consequently, as a matter of policy, they support the government and restrict themselves to reflecting official views. Second, although independent media outlets do exist, they encounter a number of problems. One is the sheer operational cost of these ventures, considering that the commercial rates (leaving aside questions of availability) for such items as equipment, premises and paper are often exceedingly high. Another is harassment. This can take a number of forms, ranging from physical assault (recent examples include in Kazakhstan the fire-bombing of the offices of an independent newspaper; and in Azerbaijan, the beating of a newspaper editor who published material sympathetic to the opposition), to indirect methods, such as pressure on printing presses not to accept opposition material, the use of the tax inspectorate to impose sudden fines, interference with distribution systems, withholding of necessary licenses and accreditation.[35]

Such pressures combine to create yet another obstacle, one that is perhaps the most effective of all: self-censorship. In order to avoid difficulties with the authorities, virtually all independent media organizations prefer to mimic the state sector in their coverage of news and information. Thus they do not provide an alternative source of information, but rather reinforce the official line. Under these circumstances, it is perhaps not surprising that the general public tends to regard the national media, state or independent, with a considerable degree of cynicism.

Broadcast and print media from abroad do penetrate these countries, but only to a limited extent. Russian television channels are quite widely watched and are generally regarded with respect. However, in Turkmenistan and Kazakhstan, the state authorities pre-record programs, then rebroadcast them: this allows for editing and censorship, when deemed necessary. In Azerbaijan, Russian and also Turkish channels can be received directly.

The British Broadcasting Corporation's World Service, Voice of America and other such foreign stations are available on radio, though not always without interference. In all three countries, cable television provides access to Western channels; however, such facilities are beyond the reach of most ordinary citizens. The same is true of the Internet: opposition groups abroad, especially Turkmen, make active use of this means of communication, but very few members of the general public within the republic have access to it.

Foreign Influences on Domestic Policies

Turkey, Iran and Russia

Foreign influences on the development of the post-Soviet political systems in Azerbaijan, Kazakhstan and Turkmenistan have been relatively small. In the immediate aftermath of the collapse of the Soviet Union, it was assumed by some foreign commentators that the region would seek to follow either the so-called "Turkish model" of Western-style democracy, or the "Iranian model" of Islamic fundamentalism. In fact, this issue was never a matter of debate in these ex-Soviet states. The main priority for both the leaders and the population at large was to preserve stability. Far from being radical attempts to abandon the Soviet political legacy, the principal innovations were cosmetic changes to the terminology of government. For a while Azerbaijan, embroiled in the war over Karabakh, did seek, under President Elchibey, a closer relationship with Turkey, but this did not result in political reforms in the domestic sphere.[36] Relations between Iran and Azerbaijan were strained at this time, since Azerbaijan regarded Iran as pro-Armenian. Also, President Elchibey favored the creation of a "greater Azerbaijan," to be achieved by the annexation of northern Iran (home to some 4.5 million ethnic Azeris).[37] Under President Aliyev, relations with Iran have improved, but are mostly restricted to trade and cultural exchanges. Turkmenistan and Kazakhstan have cordial relations with both Turkey and Iran.

There were also concerns that Russia would try to re-assert control over these countries. This has not happened in any direct sense, but undoubtedly, Russia continues to be an important factor. Proximity, coupled with sheer size of territory, population and military strength, render it an intimidating presence in regional politics, especially for neighboring Azerbaijan and Kazakhstan.[38] Moreover, Russia has many channels through which leverage can be exerted. Whether it will choose to use them will depend on the leadership in the Kremlin. Relations with President Yeltsin are cordial and he is, to some extent at least, a known quantity. Who will succeed him is inevitably a question that is of very pressing concern to the governments of Azerbaijan, Kazakhstan and Turkmenistan. Under a new Russian leader serious problems could arise.

One set of levers that Russia has at its disposal, relates to the economic sphere. The new states have made considerable strides in diversifying trading links in recent years, but nevertheless Russia remains an important economic partner.[39] Furthermore, in the development of the hydrocarbon resources of the Caspian Sea, Russia is a key player. As a littoral state it can stymie efforts to resolve the legal status of the Caspian Sea, thereby jeopardizing the aims of these new states to exploit offshore oil and gas fields. Russia also has the power to facilitate or delay the export of oil and gas from these countries across its territory.

Another set of levers relates to the Russian diaspora in these countries. Kazakhstan has some six million ethnic Russians, mostly settled in compact groups in the northeast near the Russian border. The Russian populations in Azerbaijan and Turkmenistan are far smaller (350,000 or less), but they are mainly concentrated in cities and occupy important niches in the administrative and professional structures. In the early 1990s, the Russian government seemed eager to assume the role of extra-territorial protector of these communities. This raised fears in the region that ethnic clashes might be manufactured in order to give Russia a chance to intervene. This did not happen and in recent years Russia appears to have been too preoccupied with its internal affairs to pay much attention to such matters. It is not impossible, however, that this issue could be revived in the future.

There are several other channels through which Russia could influence events in the region, including the media and personal contacts at many levels. One potentially useful card is the fact that a considerable number of political dissidents from these countries have established a base in Moscow. They do not necessarily have the support of the Russian government, but they do present the opportunity for intrigue and covert meddling.[40] This inevitably creates a certain element of distrust. For Azerbaijan, an additional reason to suspect Russia's motives is the political and material support that the latter is believed to have given the Armenians over the Karabakh issue.

Foreign Investment

Securing foreign investment in order to develop their oil and gas industries has been a policy priority for all three countries. They have looked primarily to the West for this, and have met with considerable success. Concomitants of the investment flows from the West have been aid and technical assistance programs. Donor agencies include governmental and non-governmental organizations, foundations and institutions. Many of the programs are aimed at democratization, with special emphasis on institutional capacity building, improving human rights, and establishing legal structures. Turkmenistan has been reluctant to accept training programs aimed at influencing its political development. However, several projects have been implemented in Azerbaijan and Kazakhstan, and some progress has been made in these countries in raising awareness of democracy-building issues. Nevertheless, in terms of concrete achievements, results to date have been limited. This is hardly surprising, given the short period of time that such programs have been in operation. Also, problems have sometimes arisen owing to the donors' lack of familiarity with regional conditions, as well as to an obstructive attitude on the part of the local bureaucracy. There is, too, a lack of responsiveness on the part of the general public population, who often find such undertakings of little direct relevance to their lives. In sum, the attempts of Western countries to accelerate political reform at grassroots level in these ex-Soviet republics have scarcely begun to make an impact.

International Organizations

Since independence, Azerbaijan, Kazakhstan and Turkmenistan have joined an array of international organizations. The spread of organizations reveals a desire to balance the influence of different groupings. Apart from UN organizations, the three countries have joined, amongst others, the Commonwealth of Independent States (CIS), the European Bank for Reconstruction and Development (EBRD), the Organization of Islamic Conference (OIC), the Economic Co-operation Organization (ECO) (other members of which include Pakistan, Turkey and Iran), the NATO Partnership for Peace program, and the Organisation for Security and Co-operation in Europe (OSCE). These last two organizations lay particular emphasis on democracy-building.

The OSCE has been actively involved in commenting on human rights issues in Azerbaijan, Kazakhstan and Turkmenistan, and in assessing the way in which elections are conducted in these member states. OSCE monitors were sent to observe the Kazakh parliamentary elections of 1995 and the Azeri presidential elections in 1998. In both cases, the monitors found that there were serious shortcomings.[41] No monitors were sent to the Kazakh presidential elections in January 1999, since it was clear from the outset that the contest would not be fair and open; the OSCE took the position that if they sent monitors, this in itself would be used as a way of legitimizing the elections, even if the actual verdict was negative. They did, however, send a small team of assessors; their stringent criticisms of the proceedings were subsequently made public.[42] Human rights activists within these countries welcome such public statements, but also note that they appear to have very little effect on their governments.

Threats to Social and National Cohesion

To date, Kazakhstan and Turkmenistan have been free of major civil disorders. In Azerbaijan, too, despite the continuing problems caused by the war over Karabakh, there has been a marked improvement since 1993, when terrorism was rife. Thus today, the security situation appears to be relatively stable. However, there are a number of factors that under certain conditions of stress (for example, prolonged bouts of bad weather resulting in disastrous harvests, or turmoil in neighboring states) could precipitate serious unrest. The issues that seem most likely to cause, or contribute to, instability are listed below. They are not ranked hierarchically, not only because it is impossible to assign an objective weighting to them, but also because they are interrelated and feed off one another.

Fragmentation of Society

During the Soviet period, there were many differentials in society. However, the popular perception, reinforced by the official ideology, was that there was social equality and solidarity. Since independence, rifts between different groups have begun to open up. Some of these have historical roots, such as clan-tribal divisions. There are also other cleavages along regional, ethnic, religious and urban/rural lines. Under conditions of stress, these could be activated to create hostile factions. For the present, although there are perceptible tensions, they do not represent a serious problem. A potentially more dangerous divide is that which is emerging between the "haves" and the "have-nots." A small but extremely affluent elite has emerged in these countries. Some individuals have no doubt accumulated their wealth by legitimate means, but many are suspected of having enriched themselves illicitly. They indulge in a life-style of ostentatious opulence. Typically, their clothes and cars will be expensive foreign models, while their children will be sent to private fee-paying schools, or educated abroad in prestigious Western establishments. By contrast, large sectors of the population have seen a drastic decline in their standard of living. Unemployment is rising (local observers claim that it is far higher than official estimates).[43] The situation is exacerbated by high birthrates, especially in rural areas. The social services which used to be provided free of charge during the Soviet period are now either no longer available or have become prohibitively expensive. The numbers of those falling below the poverty line are steadily increasing.[44]

Education is one of the areas that has been catastrophically affected. In poor families where there are many children – which is the rule rather than the exception in much of the region – a choice has sometimes to be made as to who should be educated; invariably it is the sons who are favored. Thus, universal education for both sexes, one of the major achievements of the Soviet regime, is rapidly being eroded. Health care, too, has suffered: there are many who can no longer afford to pay for medical treatment and as a result, health standards are falling.[45] Due to these similar downward trends in social welfare, an under-privileged class is being created. In effect, parallel societies are being created: physically they inhabit the same country, but in economic status, social organization and cultural environment they belong to different worlds. This could eventually lead to a bitter clash of cultures.

Corruption and Malpractices

Various forms of corruption – bribery, fraud, misappropriation of public property, covert "levies," nepotism, cronyism and other malpractices – are endemic in the region (as elsewhere in the former Soviet Union). Corruption existed during the Soviet period but has increased dramatically in recent years. This decline in the standards of public ethics is to some extent the product of the uncertainties of life today. Those who are in employment, at whatever level and in whatever sphere, have little expectation that they will remain at their post for long and consequently, they milk the system as much as possible while they still have access to it. This is not simply a moral issue but has practical implications for policy implementation. It subverts attempts to press forward with administrative and economic reforms. This in turn increases social divisions by giving hidden privileges to the favored few, while excluding those who do not have the right contacts and "pull."

There is scarcely any area of public life that is not affected by this evil. In the educational sphere, for example, it has undermined the credibility of academic qualifications, since it is well known that results and degrees can be bought. Such a situation breeds cynicism and lack of respect for authority. All three presidents have spoken firmly of the need to eradicate corruption; campaigns to clean up government have been launched and many senior officials and ministers have been summarily dismissed. Nevertheless, there has been little noticeable improvement. Local gossip ascribes this to the fact that the tentacles of corruption reach up to the very highest echelons of government. Whether this is so or not, it is generally acknowledged that current levels of corruption are a serious obstacle to the efficient functioning of the state in these countries.

Organized Crime

Organized criminal networks, known colloquially as "mafia,"[46] have long been active in this region, but they have mushroomed since the end of the Soviet regime. They are particularly active in Azerbaijan, but are also well entrenched in Kazakhstan. Too little is known about the situation in Turkmenistan for it to be possible to judge the strength of such groups there, but it is probable that they do exist. The main spheres of activity of these crime syndicates include illegal trafficking in high-value goods (especially drugs), protection rackets, contract killings, black marketeering, gambling rings and money laundering. Their leaders are reputed to have close links with high-ranking members of the police force and also with national and local government, thus ensuring covert official protection.

War and Secession

Azerbaijan is the only country in the Caspian region for which war and secession present immediate threats. The seven-year long conflict over Karabakh is still unresolved. The Armenian population has established *de facto* independence. The Azeris, however, continue to believe that they will eventually win back this territory. This could probably only be achieved if Azerbaijan's future oil revenues were sufficient to enable the army to be fundamentally overhauled and re-equipped. Even if this were to happen, the Armenian diaspora would very likely succeed in mobilizing international opinion to such an extent that oil companies involved in the region would find it difficult to continue working with Azerbaijan. In the wake of the collapse of the Soviet Union, other ethnic minorities (e.g. Lezghis and Talysh) also began to militate for autonomy from Azerbaijan. After Aliyev came to power, however, these movements were crushed. In Kazakhstan in the early 1990s, it seemed possible that the Russian and Cossack communities in the northeast might attempt to secede. This danger, too, has subsided as many Slavs have now emigrated; the majority of those who remain appear to prefer an accommodation with the Kazakhs, rather than confrontation. In both Kazakhstan and Azerbaijan, the possibility of fissiparous tendencies emerging among the ethnic minorities, especially those situated in border areas, cannot be excluded but there are no indications that this will happen in the near future.

Conclusions and Outlook

Azerbaijan, albeit within very limited terms of reference, has moved furthest towards establishing an open society. There is some competition of ideas rather than merely of personalities; the Azeri mass media, though by no means free, is slightly more pluralistic than elsewhere in the Caspian region. One reason for this greater level of political maturity is the experience of the Karabakh war: this brought a succession of leaders to power and diversified the forces in society. Relatively intensive exposure to Western commercial and political norms has also accelerated the process of articulating demands and accepting the need for discussion and compromise. A third, more distant spur toward democratization has been the inspiration drawn from the first era of political awakening in Azerbaijan, over a century ago (see Appendix 2). Under Soviet rule this episode was denigrated, but in the 1980s it became a source of legitimization for the nationalist movement.

Kazakhstan at first seemed the most eager of the three countries to embark on political liberalization. However, the leadership, perhaps unnerved by the enthusiasm with which Kazakh intellectuals embraced the new freedom, soon began to falter in its commitment to democratization. The dissolution of parliament in 1993, and again in 1994, signaled a steady augmentation of presidential control. Since then it has become increasingly difficult to express in public any criticism of the current regime. The media, under growing pressure to support official views, has turned to self-censorship in order to survive. Human rights activists in Kazakhstan and abroad have noted with concern, recent moves to curb the right to peaceful assembly, and also the onerous conditions placed on the formation of political groups.

In Turkmenistan, the explicit aim since independence has been to construct a political order that accords with the culture and history of the Turkmen people. Western-style political institutions are regarded as inappropriate, since they reflect different societal values and aspirations. The concept of indigenous approaches to democracy and human rights is being promoted. It is based on an idealized interpretation of Turkmen traditions. The aim is to build consensus and minimize group competition. This is of no small importance in a country where regional and clan-tribal influences continue to play an important role in public life. The idea of a genuine opposition, putting forward alternative views, is seen as unnecessary obfuscation. The president, as signified by the title *Turkmenbashi* ("Leader of the Turkmens") that has been bestowed upon him by the nation, symbolizes and personifies national unity and oneness of purpose. The external image of this society is thus one of monolithic conformity. Yet the system is not quite as rigid as it seems for there is more diversity of opinion in private than surfaces in the public domain. Whether this will be incorporated into an open political debate in the foreseeable future is impossible to tell.

The leaders of the three states are now so firmly established in office that, barring sudden and unforeseeable events (a danger which should not be underestimated),[47] they will remain in power as long as they wish, or as long as vigor and mortality permit. In making their way to the higher ranks of the Soviet system, they acquired an extraordinarily acute instinct for survival. President Aliyev, already in his mid-70s, is the most vulnerable to health problems.[48] The other two, both under 60, could have long careers ahead of them.[49] The crucial problem, however, is that the skills they learned in the pre-independence era are of limited value today in the positions in which they find themselves. Their strengths were, first, to be able to secure their own positions; and second, to organize the implementation of directives handed down from higher authority. Since the dissolution of the Soviet Union, they have continued to show an impressive ability to enhance their personal standing. In making the transition from executives to policy-makers, however, they have been less sure-footed. Initially, it seemed as though all three were intent on introducing radical reforms. In the economic sphere, they have implemented some major changes (though often inconsistently and without adequate preparation). In the political sphere, their main goal has been the maintenance of stability. They have accomplished this by preserving the most conservative aspects of the Soviet system. Given their backgrounds, and the fact that they inherited legitimacy without having to make an independent bid for power, this is perhaps not surprising. However, the result is that they have come to appear more as caretakers rather than as mold-breaking reformers such as a Mao or an Ataturk.[50]

This situation will probably continue as long as they remain at the helm, but not beyond that. After they leave the scene (however that may come about), certain issues will inevitably arise. First, there is the question of succession. The transfer of power has to be accomplished with at least a façade of legality. In the pre-colonial past, the new leader was confirmed by some nominal form of election while in the Soviet period, the choice was made in Moscow. Today, according to the constitutions of these countries, the issue ought to be decided through the ballot box. Yet given the experience of these first seven years, it seems unlikely that this procedure will present the electorate with a genuine choice of candidates.

It is more probable that the successor will be pre-selected; a vigorous political campaign will then be mounted to give the candidate validity. It is possible that the "heir" would be chosen from within the incumbent leader's family, creating embryonic presidential dynasties. There are already hints and popular rumors that the sons of the Azeri and Turkmen presidents, and the eldest daughter (in fact, stepdaughter) of the Kazakh president, are being groomed for such a role.[51] The possibility cannot be excluded that a contender from outside the family – perhaps an erstwhile member of the present leader's entourage – might seek to gain power. Such an attempt has already been made, though unsuccessfully, in Kazakhstan (by former Prime Minister Kazhegeldin). Without the blessing of the incumbent president, however, it would be difficult to build the necessary organizational base.

The next presidents will almost certainly not be as powerful as those of today, for a number of reasons. First, the post-Soviet generation of leaders will not have had the administrative and organizational preparation of the present incumbents, and consequently, will almost certainly not be able to dominate the state apparatus to the same extent. Second, the Soviet system encouraged the formation of personal patronage; Aliyev, Nazarbayev and Niyazov have all benefited from the support of the networks that they established earlier in their careers. Present conditions are inimical to the creation of such alliances, hence the next generation will not have the same staunch power base. Third, the caliber of the bureaucracy is rapidly deteriorating. Remuneration is so low that it attracts only those who cannot make a better living in any other way. The prestige of state employment has also been eroded. The law enforcement agencies, which are of crucial importance in maintaining coercive control, have been badly affected by this trend. The result is a loss of professionalism and dubious loyalties.

There are several challenges with which the new leaders will have to contend. One is the likely increase in the power of crime syndicates. Today they are not very visible, but as the experience of Azerbaijan in 1992–93 showed, as soon as the leadership weakened, the "mafia bosses" swiftly extended their reach. Private militias mushroomed and the situation rapidly descended into anarchy. Hydra-headed, it is very probable that they will re-emerge here, and elsewhere as soon as they sense a new opportunity to flex their strength. As the experience of other countries has shown, such crime syndicates could come to play a significant role in the politics of the future, by securing a weak candidate's election – and then dictating terms. This would fatally undermine a leader's ability to take rational, independent decisions.

There is no "legitimate" political opposition of any substance in these countries at present. Those who identify themselves as opponents of the current regimes are for the most part from a background and generation similar to that of the incumbent leaders and thus share a common Soviet legacy. There are a few idealists among them, but the majority offer little that is fundamentally different from current policies. Consequently, were they to come to power (which does not seem very likely), they would probably introduce new faces rather than a qualitative change in style of government. A more fundamental challenge to the authority of the new leaders could emerge from among the younger generation (roughly speaking, those born from the 1970s onwards). The bright young technocrats who go abroad for training and work experience may acquire a more independent outlook than those who have grown up within the system. Eventually, it is possible that they will press for a more transparent, accountable style of government. The emerging business community may also come to feel that it is in its best interests to have a more orderly, open political system and this, too, could be a force for democratic change.

A more dangerous and unmanageable challenge is likely to come from the alienated, deprived sections of the population. A new post-Soviet generation is growing up, many of whom are poorly educated. Unless there is a dramatic improvement in the economy they will very likely experience long-term unemployment. At present, the overwhelming majority is politically apathetic. However, if the gulf between rich and poor continues to widen – a gulf that is not only one of material wealth, but also a contrast between the "decadent" cosmopolitanism of the rich and the conservative "traditionalism" of the poor – it is very likely that this will trigger acute instability, as has happened in many other developing countries. Such conditions would favor the emergence of a populist leader preaching a radical ideology based on nationalism or religion. Given the huge quantities of arms that are circulating in the region (overspill from the wars in the Caucasus, as well as weapons acquired illicitly from poorly guarded military bases), this could very soon turn into a bloody confrontation.

External influences on the emerging political order in these new countries have to date, as commented above, been very limited. In the future, this could change, and clearer political-ideological alignments could emerge. At present, however, the situation is still too fluid for long-term predictions. All that can be done is to point to possible poles of attraction. Of these, Iran and Turkey, for reasons of proximity as well as historical and cultural ties, are likely to prove influential for Azerbaijan and Turkmenistan, though probably to a lesser extent for more distant Kazakhstan. Russia, too, will almost certainly continue to exert an influence; one or more of these three countries, either voluntarily or under compulsion, may gravitate back into the political, cultural and economic orbit of the former "center." Yet it is also possible that they will seek to develop a different set of linkages. In Azerbaijan, for example, there is a strong pro-Europe faction; in some political and intellectual circles this extends to a desire to seek membership of the European Union.[52] In Kazakhstan, although there was initial enthusiasm for stressing the country's European credentials (and some hopes of seeking membership of the European Union), this shifted first to a concept of Kazakhstan as part of Eurasia, and then as part of Asia. If China remains stable and continues its economic growth, it is possible that it will become the dominant influence in Kazakhstan.

These and other such permutations provide plentiful material for speculation. This merely underlines the fact that these new states have as yet no clearly defined orientation. As of now, they are still struggling to come to terms with the cataclysmic upheavals of the last decade. The process of transition from Soviet republics to independent statehood, far from nearing completion, has hardly commenced. Today, these young countries are stable and comparatively low in political risk. This position may be maintained and improved, but equally, it could deteriorate and the risks may escalate very rapidly. This uncertainty does not inspire confidence among foreign investors. Yet, in the absence of significant foreign investment, there is little hope that the economic situation will improve and unless this happens, and the benefits trickle down to the population at large, the political situation will become increasingly volatile.

APPENDIX 1

Basic Data

Kazakhstan

Kazakhstan, by far the largest of the newly independent Caspian littoral states, has an area of 2.7 million sq km. It lies to the northeast of the Caspian Sea, with a coastline of some 1,890 km, being bordered by the Russian Federation to the north, China to the east, Kyrgyzstan, Turkmenistan and Uzbekistan (all ex-Soviet republics) to the south. The capital was until recently Almaty, in the southeast of the country but in 1998, it was moved to Astana, which is located more centrally. Kazakhstan has a population of approximately 15 million (1999 census estimate), comprising over a hundred different peoples. The main ethnic groups are the Kazakhs, who constitute almost 50 percent (about 7.5 million) of the total population, and the Russians, who constitute some 35 percent; other groups of significant numerical size include the Ukrainians (about 6 percent), Uzbeks, Uighurs, Tatars, Koreans and Germans.

Kazakhstan has rich resources of hydrocarbons, minerals and coals. It also has good agricultural land (for grain and cattle production) and a broad manufacturing base (though this is urgently in need of modernization). The economy declined dramatically in the early 1990s and by 1995, RDP had contracted to below half the 1990 level. However, since 1996 there has been some growth. Inflation is currently below the CIS average. Kazakhstan has attracted more foreign direct investment per capita than any other CIS member. Most of this investment has been directed towards the oil industry. Some analysts predict that Kazakhstan will become one of the world's 10 largest producers of oil, early in the next century. The estimated GDP per capita is US$1,372 (1997).[53]

Turkmenistan

Turkmenistan, located along the southeast of the Caspian Sea, has an area of 488,000 sq km. It is adjoined by Kazakhstan and Uzbekistan to the north (both ex-Soviet republics), Afghanistan to the south east and Iran to the south. The capital is Ashgabat. The country has a population of some 4.7 million, of whom 77 percent are Turkmens. Other ethnic groups include Russians, Uzbeks and Kazakhs.

Turkmenistan has the world's fourth largest gas reserves. It also has rich deposits of oil and other minerals. The agricultural sector is dominated by cotton, of which Turkmenistan is one of the largest producers in the world. An economic reform program was launched in 1995, including liberalization of prices and tighter monetary and fiscal policies. There has been some improvement, with the inflation rate being reduced from 1,748 percent in 1994 to 20 percent in 1997. However, economic recovery has been hampered by difficulties with gas export routes and markets (e.g. CIS members failing to pay in full for supplies), and bouts of bad weather which have adversely affected the cotton harvest. The estimated GDP per capita is US$426 (1997).[54]

Azerbaijan

The smallest of these three states as regards size of territory is Azerbaijan, which has an area of 86,500 square kilometers. It borders the southwest of the Caspian Sea. Azerbaijan's neighbors are, the Russian Federation to the north, Iran to the south, and Georgia and Armenia to the west. The capital is Baku. Azerbaijan includes Nakhichevan, a small Azerbaijani exclave (5,500 sq. km.), which lies to the southwest and is separated from the rest of the country by a strip of Armenian territory. To the south, Nakhichevan abuts Iran, and in the west it shares a short border (about 15 kilometers) with Turkey. During the Soviet period, Azerbaijan also included Nagorno-Karabakh (4,400 sq. km.), a province situated in the southwest of the country. Since the late 1980s the Armenian population of Karabakh (as it is now more generally known) has been seeking secession from Azerbaijan. It declared its independence in 1992, but has not been recognized internationally.

The population of Azerbaijan numbers approximately 7.6 million. The Azeris (also known as Azerbaijanis) are the main ethnic group, constituting over 80 percent of the total; other groups include the Russians (about 5 percent) and small numbers of mountain peoples. Since the early 1990s, some 800,000 refugees and displaced persons have swelled the Azeri population.

Azerbaijan has extensive oil reserves. It has been a major oil producer since the middle of the last century. Since independence, several deals have been signed with foreign oil companies. In 1994, an agreement that has been called the "deal of the century" was concluded with an international consortium to develop three offshore fields, with estimated total reserves of 3 billion barrels. Azerbaijan has a well-developed industrial sector, including a downstream petroleum industry. It also has a diverse agricultural sector (fruits, cotton, tobacco etc.), which accounts for more than 25 percent of its GDP. As elsewhere in the CIS, Azerbaijan experienced a dramatic fall in output in the early 1990s; since 1996, however, this trend has been reversed and GDP has started to increase. Inflation was brought down from 1,664 percent in 1994 to less than 4 percent in 1997. There is optimism over anticipated oil revenues from the Caspian Sea. Based on this, a growth rate of 8.5 percent is projected for 1999. The estimated GDP per capita is US$506 (1997).[55]

Appendix 2

Historical Background

Kazakhstan

The territory of modern Kazakhstan was integrated into the Tsarist empire in the eighteenth century. The Kazakhs who inhabited this region followed a nomadic way of life, traveling hundreds of kilometers during their annual round of seasonal migration. They had a complex system of clan and tribal groupings, based mainly on ties of kinship. The main political formations were three "Hordes" (*zhuz*), which had their own specific territories: the Large (*ulu*) Horde controlled the south east, the Middle (*orta*) Horde the central steppe, and the Little (*kishi*) Horde the region between the Aral Sea and the Caspian. Each Horde was headed by a Khan (leader). In theory, the Khans were elected, but in practice the choice was restricted to those of a particular lineage and moreover was usually predetermined by factional interests. A Supreme Khan was chosen from amongst the three leaders. His status was that of *primus inter pares* and his main role was to encourage unity of action between the Hordes. Generally, however, the Supreme Khans were very weak, little more than figureheads.

From the mid-seventeenth century the Kazakhs suffered devastating attacks from the Oirats, a Mongol people from the east. It was primarily in response to this menace that first the Khan of the Little Horde, then of the Middle Horde, and finally of the Big Horde sought protection from the Russian crown. (However, a section of the Big Horde looked to China for help and eventually these Kazakhs, along with the eastern margin of traditional Kazakh territory, were incorporated into the Manchu empire. Approximately one million Kazakhs still live on the Chinese side of the border.) Once the Russians had established a presence in the region, they began to build up a strong administrative and military infrastructure. Subsequently, in the first half of the nineteenth century, the Khans were deposed and the entire region brought under direct Russian rule.

Colonial rule contributed to a fragmentation of Kazakh society. Members of the Kazakh aristocracy began to acquire Russian education and some joined the imperial army. They adopted an Europeanized life-style and largely lost contact with the rank-and-file nomads. Another development was the emergence of wealthy Kazakh entrepreneurs whose success derived not from privileged birth, but from innate talent. A third trend was the sedentarization of impoverished nomads who sought paid employment in the newly established factories, mines, construction projects and other industrial enterprises. The influx of thousands of Slav settlers further undermined traditional Kazakh society. Changes such as these eventually led to an awakening of a national consciousness and the first stirrings of political awareness among the aristocracy and the emerging middle class. There were two main groups: one was ideologically close to the Russian liberals of the day; the other was Marxist in orientation. A tiny number of Kazakh-language newspapers began to appear at this time, but they had very limited influence among the population at large, owing to the very low level of literacy (about 7 percent).

Kazakh resentment over Slav immigration and other aspects of colonial rule led to a major uprising in many parts of Kazakhstan in 1916. This was brutally suppressed by Tsarist troops, but before it had been entirely crushed it merged into the civil war that broke out after the 1917 Revolution between the White Army, the Red Army, foreign Interventionists and Nationalists. Soviet rule was established in 1920.[56]

Azerbaijan

The territory of modern Azerbaijan first attracted Russian interest at approximately the same period as Kazakhstan. Peter the Great regarded this region as the gateway to the East and was eager to acquire a base on the Caspian coast. In 1722–23 Russian troops occupied Derbent and Baku. Subsequently, pressure from the Ottomans – and later from the Persians (under Nadir Shah), forced them to relinquish their conquests. However, neither the Ottomans nor the Persians succeeded in establishing a permanent presence in the area and by the mid-eighteenth century, the dominant powers here were independent local rulers such as the Khans of Baku, Gandja, Karabakh, Nakhichevan and Talysh. At the end of the eighteenth century, the Russians again occupied Baku and Derbent and this time were able to consolidate their position sufficiently to expand their rule into the hinterland. The local ruling elites were rapidly deposed and the region brought under Russian military control.

As in Kazakhstan, Russian influence set in motion a number of changes in society. By the second half of the nineteenth century, Azeri scholars, writers and educationalists were seeking to modernize and reform Islam by creating a synthesis of the best elements of Muslim and Russian culture. Questions such as the introduction of the Latin script for the Azeri language in place of the Arabic script were discussed; progressive, European-style schools were opened and a vernacular theater and opera was created.

The first Azeri-language newspaper appeared in 1875. The development of the oil industry in the early twentieth century made Baku one of the largest industrial centers in the Russian empire. It attracted a multi-ethnic urban proletariat and proved fertile ground for revolutionary ideas. Stalin, one of the leaders of Bolshevik activities in Baku at that time, described the city as the revolutionary center of the Caucasus. A nationalist Azeri party, *Musavat* ("Equality") was founded in 1911. In 1918, this party succeeded in setting up an independent government in Azerbaijan. It lasted until the spring of 1920, when it was overthrown by the Bolsheviks, aided by the Red Army.[57]

Turkmenistan

The territory of modern Turkmenistan was annexed by the Tsarist empire in the last quarter of the nineteenth century. The Turkmens, formerly nomads, were by this period semi-sedentary; the majority were located in the vicinity of the oases and rivers of southern Transcaspia. They were divided into six main clan-tribal formations, each with control over a particular territory. The tribes were headed by Khans; these were supposedly elected, but in fact they were put forward by various clan factions and in general were drawn from the family of the incumbent ruler. There was frequent internecine fighting between these groups, as also between the Turkmens and their neighbors. When Russian troops invaded this region, some of the Turkmens offered a valiant resistance, notably at the battle of Geok Tepe in 1881. However, they were unable to stem the Tsarist advance.

In comparison with Kazakhstan and Azerbaijan, the Russian presence had little effect on Turkmen society. Traditional social structures were weakened, but not destroyed. The influence of European culture was almost entirely restricted to the environs of the small Russian military settlements. In 1916 there was some spillover from the uprising in Kazakhstan, and in 1917 civil war broke out. A nationalist faction, aided by a British mission from Meshed, briefly succeeded in establishing an independent state. However, this soon fell and the Red Army took possession of the region.

Soviet Era

In 1924, some four years after Soviet rule had been established, the territory of Central Asia was divided into five main administrative units. One of these was the Turkmen Soviet Socialist Republic (SSR), another (created somewhat later) the Kazakh SSR. On the other side of the Caspian Sea, in trans-Caucasus, new administrative units were likewise being formed, among them the Azerbaijan SSR (created in 1920). The borders that were drawn in the 1920s were internal, administrative demarcations within the larger territorial entity of the Soviet Union. The only international frontier at that time was that which separated the Soviet Union from surrounding states – Iran, Afghanistan and China. After the collapse of the Union, the internal borders between the Soviet republics (which had remained largely unchanged throughout the Soviet period) became international borders between the now independent states.

Under Soviet rule, a radical and far-reaching program of modernization was undertaken. This included mass education (literacy levels were raised to virtually 100 percent), the emancipation of women (including abandoning the veil in areas where it was customarily worn), secularization, collectivization of agriculture, settlement of the nomads (completing the process that had started in the late nineteenth century) and industrialization. Modern bureaucracies were created, along with modern systems of transport and communication.

Slav immigration (mainly Russian) into the region had commenced in the nineteenth century, but it increased vastly during the years of Soviet rule. Kazakhstan was most severely affected, with the combined Russian and Ukrainian population rising from approximately two million in 1926 to nearly seven million in 1979 (or from 20 percent to almost 47 percent of the total population). Representatives of many other ethnic groups also came to these republics at this time. Some were volunteers, others political deportees; the majority of the latter were sent to Kazakhstan, with a much smaller proportion going to Turkmenistan and the other Central Asian republics.

Throughout the Soviet period there was intensive political indoctrination of the population. The official ideology of the regime,

Marxism-Leninism, was inculcated by both formal and informal means. The educational system, from nursery school upward, was saturated with ideological content. The message was continually reinforced in the media, also through the arts (paintings, plays, operas and films etc.). Sociopolitical organizations amplified this ideological work. These included the Young Octoberist and Young Pioneer movements (for 7–14-year-olds), and the Young Communist League (for 14–28-year-olds), as well as trade and professional unions.

The core of the political system, underpinning the state apparatus, was the Communist Party. The organization of the Party was highly centralized. Grass-roots control and mobilization were carried out through local "primary party organizations" These in turn were responsible to a hierarchy of administrative-geographic levels of organization (regional, provincial etc.). Each republic had its own Communist Party (e.g. the Communist Party of Azerbaijan), headed by a First Party Secretary. These republican institutions were in turn affiliated to the Communist Party of the Soviet Union (CPSU). The head of the CPSU was the General Secretary, who in real terms, was the most powerful individual in the Soviet Union. The central organs of the CPSU were situated in Moscow. Members were drawn from all over the Soviet Union, though there was always a great preponderance of Slavs. The Politburo of the Central Committee of the CPSU was the body that exercised power over the day-to-day decision making process. It convened under the chairmanship of the General Secretary and acted somewhat like a parliamentary cabinet. The numbers of members varied, with up to about fifteen full members and some eight non-voting members.

Membership of the Communist Party was regarded as an honor and admission procedures were long and painstaking. For successful applicants, there were privileges as well as obligations. Party membership was all but indispensable for advancement in the state apparatus, the armed forces or the other professions. Consequently, the large native elites that were created during this period in all spheres – political, administrative, military, intellectual, sport and the arts – were overwhelmingly members of the Party. They were not necessarily actively committed to the doctrines of the Party, but, as members of an exclusive institution, they shared a common ethic. Also, bonds of friendship and loyalty were created that often cut across other ties such as those of ethnic origin.

When the Soviet Union disintegrated, the Communist Parties of the newly independent states were either renamed, as in Turkmenistan, or re-created on a much smaller scale, as in Kazakhstan and Azerbaijan. There are virtually no government officials in these countries that would still claim to be Communists. However, the overwhelming majority of those who are over 40 years of age are ex-Party members and in their attitudes and outlook, the imprint of the Party experience is still very strong. Also, many of the informal networks and alliances of today are founded on links formed within the ranks of the Party. Thus, though the Communist Party no longer has any real power in these countries, nevertheless, it continues to exert an indirect influence on the way in which they function.

5

The Industry's Race for Caspian Oil Reserves

Julia Nanay

Introduction

Why have so many oil and gas companies been trying to get access to the oil and gas reserves in the Caspian region? Despite the depressed price environment for oil and gas, new companies are still coming. Will their interest continue, will it intensify, or will it begin to diminish?

In terms of the broader picture for the oil industry, a number of issues today define the business imperatives of oil companies. To start with, today's big oil companies understand that they are not successful *because* they are big. They are big because they have been successful *in the past.* Big oil companies want to stay big, and they are trying to become even bigger. How can they do that? There are two ways: through merging with or acquiring other companies and/or growing through exploration/production or development deals. Both of these avenues are increasingly being applied as companies find reserves in the Caspian Basin.

Search for New Investment Opportunities

The decade of the 1990s has been redefining the business environment for the oil and gas industry. It began with the opening of Russia to foreign investment, especially its huge West Siberian and Sakhalin Island oil fields. This occurred at a time when production from the last great oil provinces of the Alaskan North Slope and the North Sea was declining. Companies were searching for new international opportunities. The Middle East was largely off limits, and the isolation of Iraq was reinforced after the Gulf War. Few areas remained that offered investment opportunities of any size or scale – certainly not comparable to Russia. Thus, the first rush to Russia was in 1992–93. That euphoria was subsequently tempered by the lack of progress on concluding Production Sharing Contracts (PSCs). By 1994, attention had shifted to Kazakhstan and Azerbaijan. This was a period (1993–94) where the competitive field in terms of numbers and types of companies was expanding, putting enormous pressure on all companies – big and small – to stake out new growth opportunities. The Caspian Basin held out that promise for growth.

Expanding the Competitive Arena

From the standpoint of competition in the oil industry, since 1993, the Russian oil industry has been transformed with the creation of a series of integrated firms. By 1997, nine Russian oil companies had joined the ranks of the world's 50 largest oil companies. The *Petroleum Intelligence Weekly*'s (*PIW*'s) list of top 50 firms in December 1997 included: Surgutneftegaz (#23), Sidanco (#26), Yukos (#28), Lukoil (#29), Gazprom (#31), Rosneft (#39), Sibneft (#40), Tyumen Oil (#42), and Slavneft (#44).[1] These firms gained sufficient clout in Russia to block the entry of Western firms into major projects in Western Siberia. Of these firms, Lukoil and Rosneft now have diverse interests in the Caspian region and have been joined there by another Russian firm, Central Fuel Company, which is not currently on the Top 50 list.

In 1997, in addition to the nine Russian companies on the *PIW* Top 50 list, there were 21 state-owned or national oil companies. National oil companies asserted themselves on their own home turfs and, in some cases, on the world scene. At the same time, while the opportunities worldwide for investments of a certain size and return were shrinking, many of these companies were at the heels of the majors and independents seeking access to the same opportunities. With no shortage of capital to invest internationally, but with a shortage of opportunities, companies "rushed" to the Caspian.

The Rush to Azerbaijan and Kazakhstan

Between 1994 and 1998, 26 companies from 13 countries signed PSCs in Azerbaijan. Kazakhstan could tell a similar story, but the structure of its oil industry is very different. In addition to the majors and other well-known international oil companies, it has attracted some different types of players. The major acreage holder in Kazakhstan, First International Oil Company (FIOC), was created by Kazakh and Western oilmen to take advantage of the huge resources and upstream opportunities in that country. FIOC holds 33,600 square kilometers in Kazakhstan, of which it has farmed out 11,800 sq. km. to Amerada Hess.

New Pressures

The competitive field in the oil industry in general is now beginning to contract and in the Caspian, it should enter a period of consolidation – despite recent new entrants, like US companies Amerada Hess and Phillips. In fact, companies are increasingly moving towards cooperating with each other and sharing information in daily operations rather than fighting for access. But there is new pressure brewing as some companies have over-extended themselves in the Caspian and worry about meeting their capital commitments in the region.

The Azerbaijan International Operating Company (AIOC), which signed its PSC in September 1994, has spent close to US$2 billion to date and incurs US$1 million a day in costs to keep its operations going, while producing about 90,000 b/d. Constant cost overruns have plagued the consortium over the last three years, making some companies worry about whether they can afford to maintain their positions. The AIOC partners are BP-Amoco, Unocal, Exxon Mobil, Devon Energy, Statoil, Ramco, Lukoil, Itochu, TPAO, the Delta–Hess joint venture and SOCAR.

Some companies may be forced to cut down operations. Other companies may be eliminated because of a general trend in the industry toward mergers and acquisitions. In fact, mergers and acquisitions will surely change the face of Caspian competition over the next year. It will change it so that there are fewer companies competing, maybe even different companies.

Returning to the premise that companies are big because they have been successful in the past, this is true for all of the companies on the *PIW* Top 50 list, both private and state-owned. Most troubling for the future of all of these companies is how they can continue to grow and to be profitable in the medium term, particularly with a US$12–14 per barrel oil price environment. The burst of activity of the early and mid-1990s is being decimated. The competitive field is narrowing. Some of the companies on the Top 50 list will be eliminated and absorbed into other firms, as companies achieve growth through mergers and acquisitions.

But while much of the news is bad and capital availability is tightening, there is a glimmer of hope on the horizon. New investment opportunities may be opening. This could happen with respect to Russia and it is certainly occurring in the Middle East. Of course, these opportunities could draw investment away from the Caspian. Over the last two years, companies have been dealing with some of the most difficult risks in the Caspian, which are discussed below. These risks could begin to temper the interest of some of the smaller players. Certainly, the big companies are there for the long term. Resolving the exit route dilemma for oil and gas exports has been just one of the sources of frustration for all of the companies operating in this region, particularly the smaller firms, which are under greater pressure to show results. Smaller firms do not have the breadth and depth of international portfolio as the larger firms. For the latter, delays or disappointments in the Caspian can be offset by successes in other areas.

Which Countries Dominate in the Caspian?

As the industry consolidates in the Caspian, the balance of countries with controlling interests may shift away from the United States to others. This could have implications for exit route decisions. Until the BP–Amoco merger, US companies dominated in the Caspian. The United States was the outside player with the most commercial stake, which the US government has been translating into political clout. The balance in the Caspian as a whole is now more evenly divided between the United States and the United Kingdom, but in Azerbaijan, the United Kingdom now takes the lead. Russia is the third principal political player in the region, with growing commercial interests.

Russia moved into the Caspian upon the heels of foreign companies in the early to mid-1990s. When foreign oil and gas companies began to shift their attention away from Western Siberia toward the Caspian in 1993–94, Russian company Lukoil followed suit and staked out important upstream and downstream positions in Azerbaijan and Kazakhstan.

Lukoil, the Russian company with the most extensive interests in the Caspian, wanted to be present in the two countries which have attracted the most investment dollars from the West.

Why are Companies in the Caspian Basin?

Countries are not flocking to the Caspian just because of the size of the proved reserve base or even anticipated production, since these are not particularly significant in comparison to some other areas of the world.

FIGURE 5.1
Oil Reserves, 1997

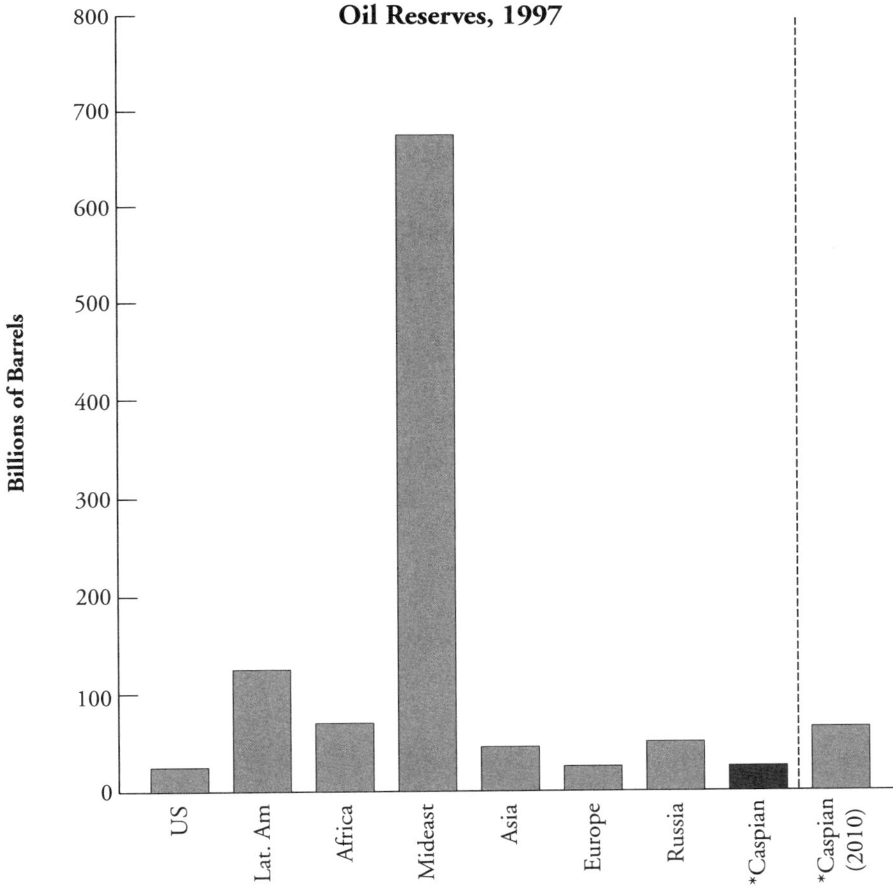

Note: *Caspian: Azerbaijan, Kazakhstan, Turkmenistan
Source: The Petroleum Finance Company Ltd

FIGURE 5.2
Oil Production, 1998

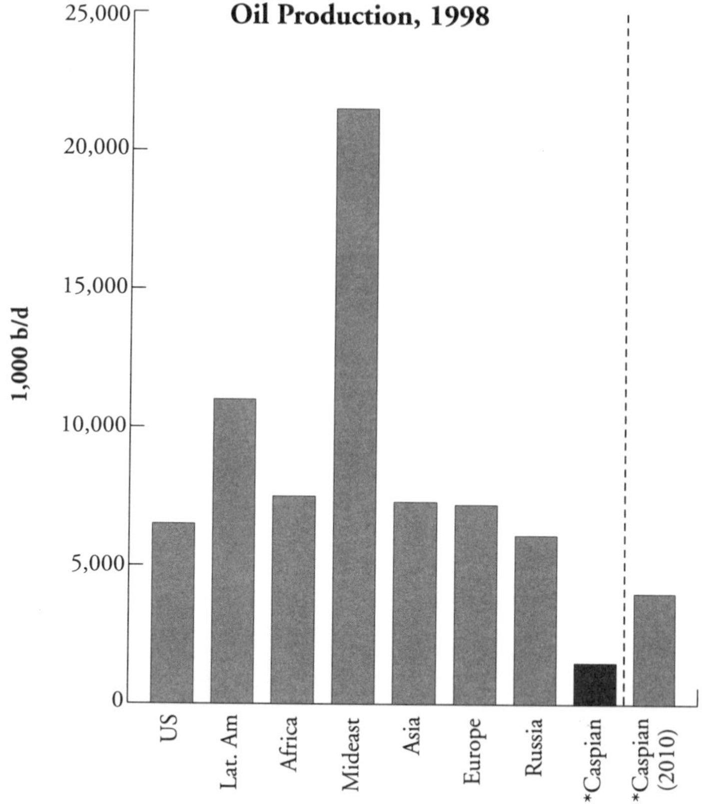

Note: *Caspian: Azerbaijan, Kazakhstan, Turkmenistan
Source: The Petroleum Finance Company Ltd

FIGURE 5.3
Gas Reserves, 1997

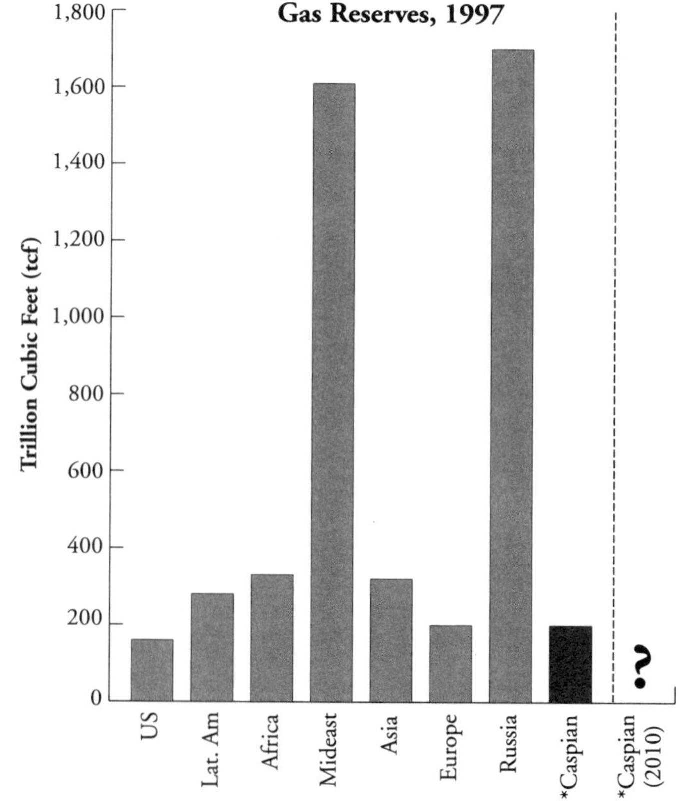

Note: *Caspian: Azerbaijan, Kazakhstan, Turkmenistan
Source: The Petroleum Finance Company Ltd

FIGURE 5.4
Gas Production, 1997

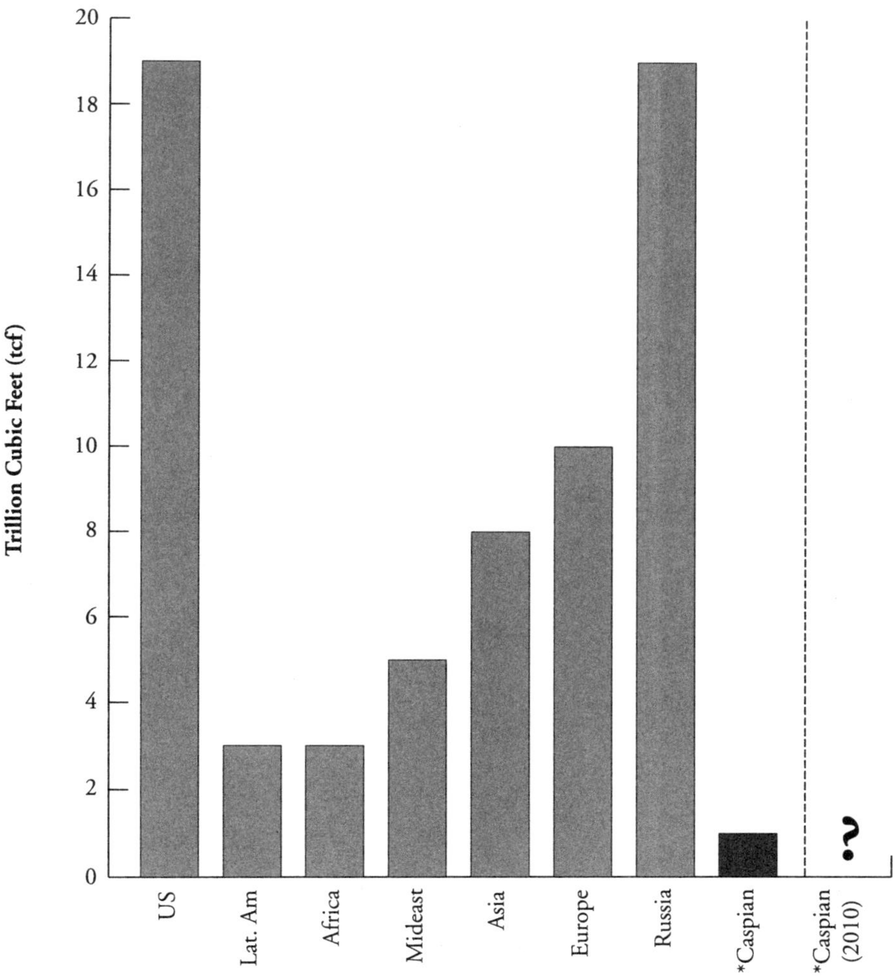

Note: *Caspian: Azerbaijan, Kazakhstan, Turkmenistan
Source: The Petroleum Finance Company Ltd

Companies rushed to invest in Azerbaijan, Kazakhstan and Turkmenistan because there is limited or no access to oil and gas reserves and production in many of the most prospective countries of the world. In other words, the world offers limited opportunities for oil companies to grow; without access to new reserve and production sources, companies are unable to grow.

FIGURE 5.5
Access to Proven Reserves

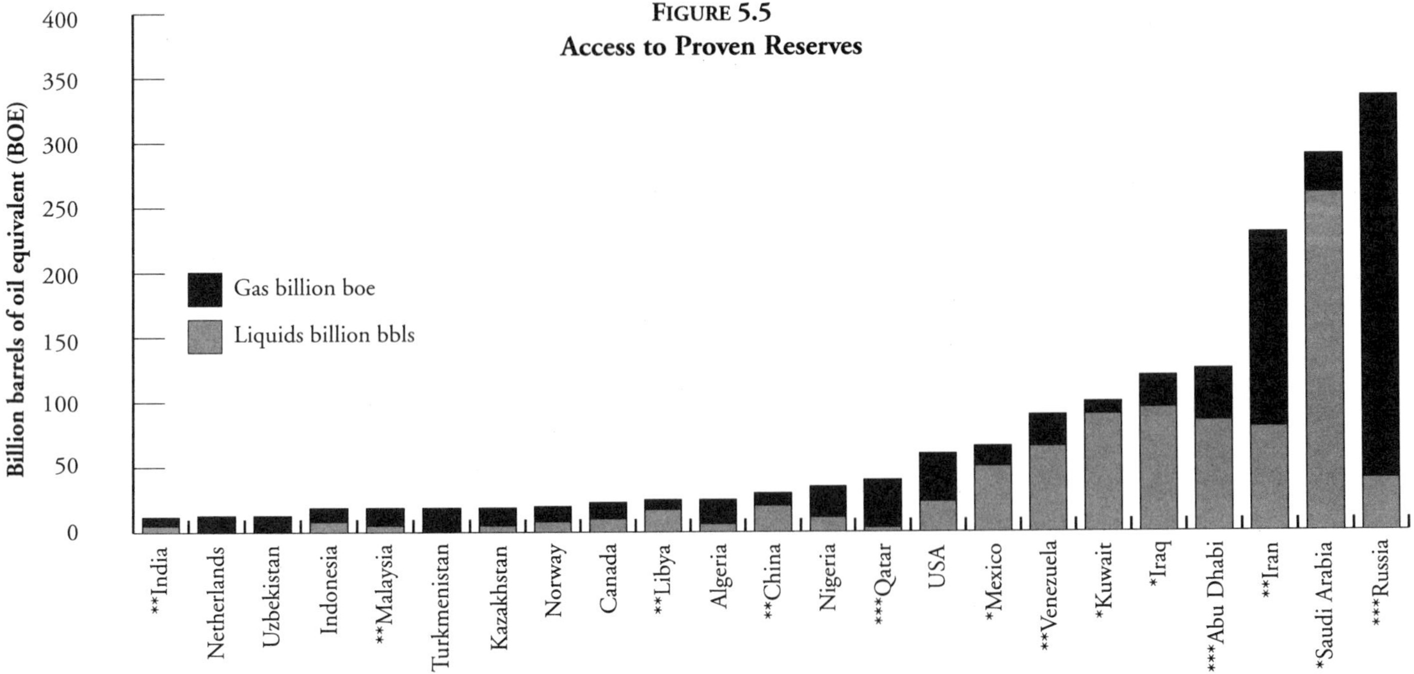

Notes: *Foreign firms have no access to reserves and production
**Foreign firms have limited access
***Foreign firms can book oil and condensate as liquids reserves; gas reserves revert to government

Source: *BP Statistical Review*, 1996

- Enormous crude oil and gas reserves in countries where there is little to no foreign firm opportunity or access
- At the same time, foreign firm access is biased toward gas dominated basins
- Saudi Arabia and Iraq control 35% of the world's oil reserves; Russia (34%) and Iran (15%) control just over 49% of the world's gas reserves. Their gas reserves exceed their oil reserves

In 1998, this was changing, as Iran was being pried open and Kuwait moved closer to allowing foreign upstream investments, while sanctions on Libya were suddenly also closer to being lifted.

FIGURE 5.6
The Top 15 in Oil Reserves, 1997

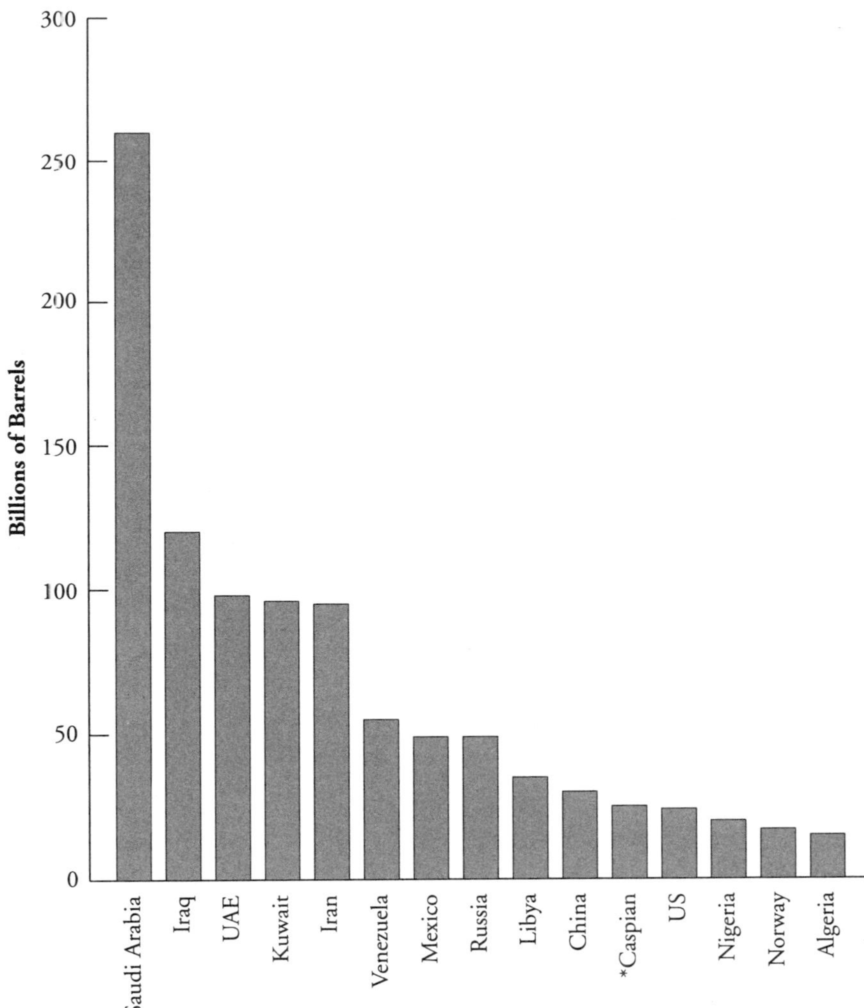

Note: *Caspian refers to Azerbaijan, Kazakhstan, Turkmenistan
Source: The Petroleum Finance Company Ltd

Figure 5.7
The Top 15 Gas Producers, 1997

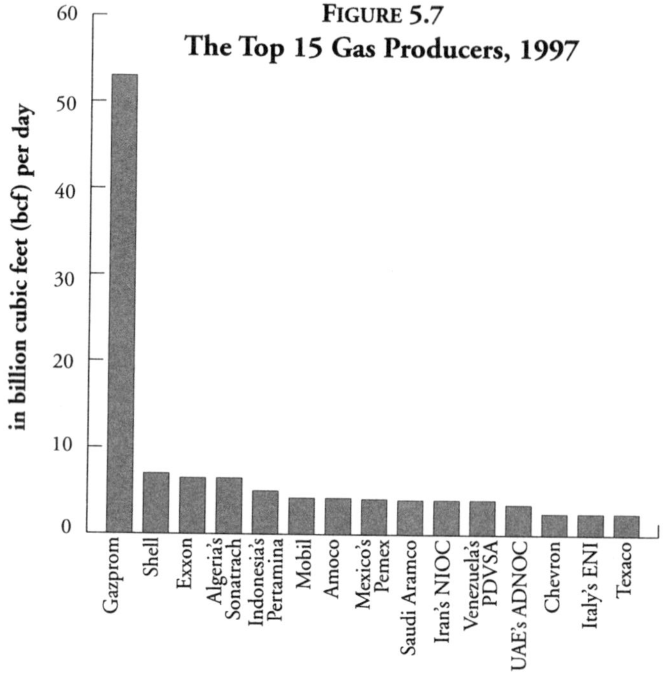

Marketed Production after subtracting reinjection, flaring & other losses from Gross Production

Note: Russia, with 34% of the world's gas reserves, accounts for 25% of the world's marketed gas production. Iran has 15% of world reserves and accounts for 1.7% of its marketed production

Source: Cedigaz, Natural Gas in the World, 1997 Survey; converted from bcm to billion cubic feet per day and PIW, December 22, 1997, Special Supplement on World's Top 50 Companies

Figure 5.8
The Top 20 Gas Reserves, 1997

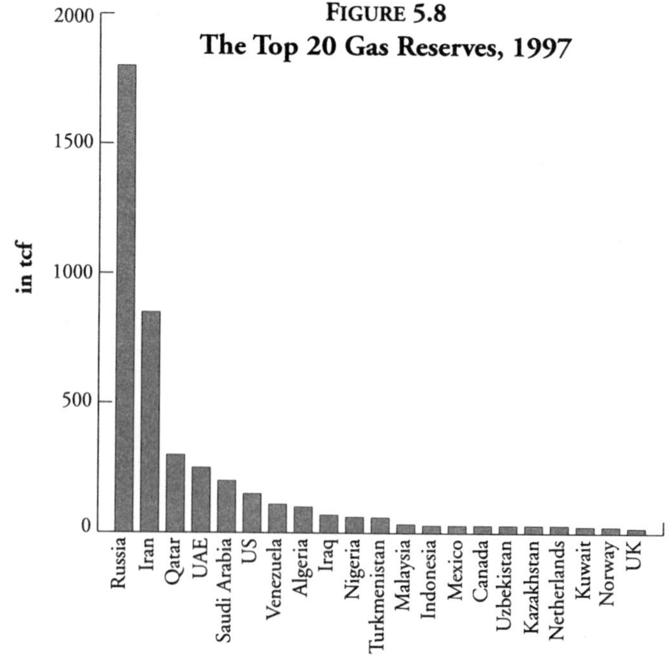

Note: Russia is the world's largest gas reserve holder, with 1,700 tcf, of which Gazprom controls only 69%. Iran has the world's second largest gas reserves or 815 tcf, and accounts for about 47% of the gas reserves of the Middle East and for about 15% of the world's total gas reserves

Source: Cedigaz, 1997

Energy Corporations Continue the Search for New Core Producing Areas

Getting back to the oil and gas exploration and production business, what does all this mean in the context of the Caspian? Looking at what has made successful companies big companies, it is their core producing areas. Central to any successful upstream business segment is the establishment of one or more core producing areas.

Examples of core producing areas include the North Slope of Alaska for BP and Arco; the North Sea for BP, Elf and many others; West Africa for Elf and Chevron; Indonesia for Exxon Mobil, Total, Unocal and others; and Thailand for Unocal and others. Core producing areas provide sizeable reserves and production. This is where firms make the most profits. These core areas provide continuous business opportunities. In the case of Unocal, for example, Thailand represents a very successful core producing area.

However, looking at the future production profiles of most companies going out after the year 2005, the same story appears again and again: existing core producing areas are becoming old and production is declining. Companies are seeking new core producing areas. The Caspian Basin is one of the only regions – free of US or multilateral sanctions and where access is relatively unimpeded by governments – which offers companies the scale of opportunities for investment that could allow them to create new core producing areas.[2] Furthermore, companies can book the oil and gas reserves in this region, which is then reflected in their stock price and is seen as bringing value to their shareholders. Another important attraction is that companies can operate under PSCs.

For BP–Amoco in Azerbaijan and Chevron in Kazakhstan, these countries provide continuing business opportunities in what can be considered a core producing area for each company. These companies can keep proving additional reserves in these core areas and can keep producing more oil and gas so that they can move up the learning curve, realize economies of scale and achieve stability in production and cash flow. In an effort to increase further their Caspian operations, BP-Amoco has extended its upstream activities to Kazakhstan, while Chevron has diversified to Azerbaijan.

Azerbaijan: BP-Amoco Is the Lead Player

Projecting into the future, Azerbaijan may very well represent an important core area for BP-Amoco. After all, with the merger of these two companies, 5 of the 13 offshore PSCs are now operated by this new entity, including the contract for AIOC, where BP-Amoco alone now has a 34 percent controlling stake and claims 1.4 billion barrels of oil reserves. Add the eventual reserves in the other 4 PSCs and Azerbaijan could prove to be a company-maker for BP-Amoco.

Azerbaijan: The Risks

It is important to note that AIOC has the only project in Azerbaijan with proven reserves. Everything else is speculative, although announcements with respect to BP-Amoco/Statoil's Shah Deniz project are due soon. Already there is some disappointment because the second PSC for the Karabakh field (the CIPCO project, operated by Pennzoil with Lukoil and Agip as partners) has now drilled three wells at a cost of about $100 million, which show no commercial volumes of oil or gas. The fourth PSC, BP-Amoco's NAOC project is likely to reinforce this trend. The third PSC, BP-Amoco/Statoil's Shah Deniz may contain more gas than oil. If increasing amounts of gas are found in Azerbaijan, this will present a dilemma for the foreign companies and for Azerbaijan. Squeezed in between two major gas reserve holders and producers – Russia and Iran – the competition will be tough. Marketing this gas would prove to be a much bigger headache than marketing the oil. In addition, the gas infrastructure – gathering systems, pipelines, processing facilities – will be costly to develop. Some companies do not want to take the financial risks involved in gas developments. While AIOC could also produce significant volumes of gas, this gas is not part of the Western companies' PSC. It reverts back to SOCAR, Azerbaijan's state oil company.

The CIPCO partners have written off the costs of drilling in the Karabakh field. For a company like Pennzoil (now Devon Energy) one of the early entrants to Azerbaijan, this is a big disappointment. Pennzoil sold down a portion of its stake in AIOC to be able to finance its CIPCO position. It was counting on Karabakh to be a big winner. So was Azerbaijan, because the success of Karabakh would have made a Baku-Ceyhan oil pipeline commercially feasible. The lack of oil in

Karabakh changed the whole psychology of the offshore play in both Azerbaijan and Kazakhstan. It burst the bubble of excitement over the Caspian, as companies woke up to the reality that these offshore projects were true exploration plays, with no guarantee of finding oil. The next structure, currently being drilled in Azerbaijan, Shah Deniz, will be an indicator as to whether the trend is toward more oil, more gas, or even dry holes in the Azeri section of the Caspian. The uncertainty for Azerbaijan is that based on recent drilling results, the country could be more gas prone in the near-term than oil prone – though drilling in later PSCs may reverse this trend.

Implications for US Supported East–West Routes

The dilemma for the US government is that in November 1997, when it first began making its enormous push for East–West pipeline routes, the euphoria over first production from AIOC got in the way of rational thinking. It was just taken for granted that the US vision of billions of barrels of oil and gas reserves in the Caspian Basin justified the construction of a multibillion dollar pipeline project based on what the US considered best for its interests and hence, also best for the Caspian countries. With the commercial feasibility of such a project now unraveling, the US cannot let go of its plans because it has exerted too much political capital on a Eurasian transport corridor, goaded too many countries into believing it would become a reality, and is determined not to lose face. But how do you get a multi-billion dollar pipeline built when oil prices have hit such low levels. In real terms, oil prices are lower today than they were at the beginning of World War II. This is the other part of the equation that the US could not factor in back in November 1997. Even if oil prices firm, oil companies will adjust their planning to test all projects at much lower price levels than previously, in order to be prepared for a repeat of 1998.

Kazakhstan: Chevron Is the Lead Player

Other than AIOC, in terms of having actual oil production on hand, the major success story in the Caspian is Kazakhstan's TengizChevroil joint venture. In this one project alone, US company Chevron has over 2 billion barrels of liquid reserves. The TengizChevroil oil and gas field "production or development deal" onshore northwest Kazakhstan (Chevron/Exxon Mobil/Arco[LukArco], 72 percent), represents the Former Soviet Union's (FSU's) most significant PSC to date. This remains under majority US control. The PSC was signed in September of 1993 and in 1999 is expected to produce about 180,000 b/d (this is the largest amount produced by any foreign-partnered project in the FSU). Reserves in the Tengiz field are estimated at between 6 and 9 billion barrels of liquids and over 10 trillion cubic feet (tcf) of gas. Per barrel lifting costs are now under US$2 per barrel. As Chevron succeeds in solving the exit route dilemma by securing numerous different exit options and moves forward on the construction of the Caspian Pipeline Consortium (CPC) pipeline,[3] it may be well-positioned to add more assets in Kazakhstan. Chevron has begun to drill new exploration wells at Tengiz with the expectation that more reserves will be added. While BP-Amoco is the key company in Azerbaijan, Chevron is the most significant investor to date in Kazakhstan.

Kazakhstan: The Risks

As with the offshore situation in Azerbaijan, there is similar uncertainty in the offshore Caspian in Kazakhstan, where drilling operations on the first well began in October 1999. The Offshore Kazakhstan International Operating Company (OKIOC), comprised of nine companies [Shell, BP/Statoil, BG, Agip, Total, Mobil (now Exxon Mobil), as well as Phillips and Inpex], conducted three years of seismic exploration from 1993 to 1996, and signed that country's first offshore PSC in November 1997 during President Nazarbayev's visit to Washington, DC. The Kazakh offshore project requires extensive exploration and while the area is deemed extremely promising, the actual size of oil and gas reserves cannot be assessed until drilling has progressed beyond the first well. After the first well, consortium members will know if the acreage holds oil or gas or neither. It will be many months before they can estimate actual volumes that can be produced. This is the project that the United States was looking to as a possible fill for the Trans-Caspian portion of the Baku–Ceyhan (East–West) pipeline but, obviously, there is no way to judge what volumes will be available yet and most definitely, the necessary oil (upwards of 500,000 b/d) will not be there until after 2005 (if at all).

Exxon Mobil: The Most Extensive Regional Presence

The US company with the most extensive presence in the Caspian is Exxon Mobil. This company farmed into the Kazakh government's share of Tengiz in 1996, taking 25 percent. In addition, Exxon Mobil has an exploration area near Tengiz, Tulpar Munai, where it has brought in Shell and a number of Japanese companies as partners. It is the only US company in the offshore consortium in the northern Caspian/Kazakhstan, where it signed a PSC in November 1997 during President Nazarbayev's US visit. In August 1997, Exxon Mobil took its first offshore position in Azerbaijan. In Turkmenistan, Exxon Mobil is partnered with the UK's Lasmo in two PSCs, which cover most of the western oil-prone area of the country. Mobil remains the only foreign company with upstream interests in all three countries – Azerbaijan, Kazakhstan and Turkmenistan – having merged with Exxon which had much smaller stakes across the board. Exxon Mobil is also present in Russia – in Sakhalin. Among US companies, Chevron, with Kazakhstan's Tengiz field, is the most concentrated on a single project in a single country. Chevron took its second regional position in August 1997 in Azerbaijan.

Substituting the North Sea

The real problem for the development of the oil and gas industry is that no major oil-producing area has emerged through exploration in more than 25 years since the North Sea was discovered. Companies are seeking to replace the North Sea. And, we will keep hearing that reserves in the Caspian Basin could be the equivalent of several North Seas. Of course, while the reserves are projected to be large, it is difficult to know until actual drilling occurs whether they will contain oil or gas and/or, in some cases, companies may end up drilling dry holes. The verdict is still inconclusive as to whether the Caspian Basin will be the next North Sea or even many North Seas.

Kazakhstan could prove to be the most important oil story in this region, in terms of the likely size of its resource base (at least onshore). Kazakhstan's oil reserves (8–10 billion barrels of proven reserves) are probably underestimated. The Tengiz field alone contains 6–9 billion barrels of oil reserves and upwards of 10 tcf of gas. New drilling at Tengiz this year could prove more reserves. Unlike Azerbaijan, all of Kazakhstan's existing oil and gas production is onshore, which until now has not exposed it to the battles being fought over the Caspian's legal status. Furthermore, it has sizeable remaining oil and gas reserves onshore. Estimated onshore oil reserves are between 15 and 30 billion barrels. Offshore, the potential could be much higher. In terms of proven reserves, Kazakhstan falls below Norway (11 billion barrels), but its estimated reserves make it the equivalent of two or three Norways. What is most striking about Kazakhstan is that its proven gas reserves of near 2 trillion cubic meters (64 tcf) are also enormous (well in excess of Norway's 48 tcf). With production of about 515,000 b/d in 1997, Kazakhstan's oil output is but a fraction of Norway's 3 million b/d. Gas production of about 6 billion cubic meters a year (211 billion cubic feet/year or 580 million cubic feet/day) is also a fraction of Norway's 41 billion cubic meters/year (1.4 trillion cubic feet/year). Kazakhstan exports about 200,000 b/d of crude oil. This is a country with huge resources.

If companies stay the course and maintain their commitments in Azerbaijan, Kazakhstan and Turkmenistan, this is a region that definitely has the potential to be the next North Sea, absorbing the capital of independents, majors and national oil companies and forming an important new axis in worldwide energy supply.

In terms of replicating the way the industry has evolved in the North Sea, Azerbaijan provides the best showcase, with 14 countries and 25 companies – small and large – represented offshore in 13 PSCs (although two of these are being terminated). Onshore, there are another 3 PSCs, bringing the total number of PSCs to 16 and the total number of companies working in Azerbaijan to 28. This provides a fascinating microcosm for how the oil and gas business could develop throughout the region, with one caveat – the larger companies may increasingly dominate as smaller companies either sell off their interests or are acquired. Companies need long staying power (i.e. deep pockets) to be able to absorb the risks of operating in the Caspian in a depressed oil price environment.

6

The Development of Caspian Sea Oil: Implications for OPEC

Wilfrid L. Kohl

There is considerable uncertainty regarding the amount of oil reserves in the former Soviet republics in the Caspian Sea region. This chapter reviews the range of recent estimates for Caspian oil reserves and production and export potential. It then explores some implications for OPEC and the world oil market. It concludes with comments on US policy toward Caspian Sea oil development and energy security.

Reserves

Proven oil reserves are those quantities of oil that can be reasonably recovered under current economic and technological conditions based on geologic and engineering evidence. Using this definition, the Energy Information Administration (EIA) of the US Department of Energy (DOE), citing data from Petroconsultants (internationally respected energy consultants), lists proven oil reserves in the Caspian region totaling 32.5 billion barrels, while potential reserves (a more speculative number based on seismic data but not actual drilling) could add another 186 billion barrels, bringing total reserves to well over 200 billion barrels (Table 6.1).[1]

In its 1998 study, *Caspian Oil and Gas*, the International Energy Agency (IEA) estimates proven oil reserves in the range of 15–40 billion barrels (based on a survey of other studies) and possible additional reserves in the range of 70–150 billion barrels.[2]

TABLE 6.1
Oil Resources in the Caspian Sea Region
(Billion Barrels)

Country	Proven Reserves	Potential Resources	Total Resources
Azerbaijan	12.5	32	45
Iran	0.1	15	15
Kazakhstan	17.6	92	110
Russia	0.3	7	7
Turkmenistan	1.7	38	40
Uzbekistan	0.3	2	2
Total	**32.5**	**186**	**218**

Source: "Destiny" International Energy Forecast Software (Houston, TX: Petroconsultants, 1998). As reprinted in EIA, *International Energy Outlook,* 1998.

How do these reserve figures compare with other oil-producing regions? The 1998 edition of the *BP Statistical Review of World Energy*, which uses conservative figures for proven reserves, lists Caspian proven reserves at about 16 billion barrels.[3] For comparison, proven reserves of other major producers are reproduced in Table 6.2.

Taking the lower BP estimate for Caspian proven reserves at 16 billion barrels, this represents about 1.5 percent of world proven reserves in 1997. Adding higher-end potential reserve estimates bringing Caspian reserves up to 200 billion barrels would yield a figure of approximately 16 percent of a larger amount of total world reserves. This is of course a highly speculative number, but if such reserves became proved, the Caspian potential would still be less than one-third of total Middle East proven reserves (677 billion barrels according to BP, or 65 percent of world reserves at the end of 1997). At 100 billion barrels (still an optimistic estimate) Caspian proven reserves would represent about 9 percent of world proven reserves. Either higher number would represent a substantial addition to non-OPEC proven reserves, currently running around 175 billion barrels.

In certain areas around the Caspian, especially Azerbaijan, oil and gas production has had a long history. However, in other areas only limited drilling and seismic study have been done in the offshore areas, and only limited onshore deep drilling has been accomplished thus far. Of

TABLE 6.2
Proven Oil Reserves of Major Oil Producers, 1997
(Billion Barrels)

Country/Region	Reserves
Saudi Arabia	261.5
Iraq	112.5
UAE	97.8
Kuwait	96.5
Iran	93.0
(Others incl. Oman, Qatar, Syria, Yemen)	15.6
Total Middle East	**676.9**
Mexico	40.0
USA	29.8
Canada	6.8
Total North America	**76.6**
Total North Sea (Denmark, Norway, UK)	**16.3**
Venezuela	71.7
Brazil	4.8
Colombia	2.8
Argentina	2.6
Ecuador	2.1
(Others)	2.2
Total South and Central America	**86.2**
Libya	29.5
Nigeria	16.8
Algeria	9.2
Angola	5.4
Egypt	3.8
Gabon	2.5
(Others)	2.8
Total Africa	**70.0**
Total Russian Federation	**48.6**
Total Caspian Republics	**16.0**

Note: Figures include only major oil producers, not total oil in each region.
Source: *BP Statistical Review of World Energy*, 1998.

the three distinct geologic basins underlying the Caspian, the Northern basin is geologically the most promising, followed by the Southern basin and then by the middle Caspian. Gregory Ulmishek, who is working on a new report for the US Geological Survey due to have been published in 1999, estimates potential reserves at 25 billion barrels in the South Caspian, between 4 and 5 billion barrels in the middle Caspian, and 75 billion barrels in the North Caspian, making a total of 105 billion barrels.[4]

Much of the Caspian region remains to be explored. But it is clear that promising oil reserves are present in Kazakhstan and Azerbaijan, and to a lesser extent in Turkmenistan and Uzbekistan. More important than estimating the exact amount of reserves is the attraction of foreign investment to expedite the exploration and production of oil and the establishment of transportation infrastructure (such as pipelines) to transport the oil from this landlocked region.

Production and Export Potential

Both Kazakhstan and Azerbaijan have created favorable investment climates for foreign oil companies. All four of the major republics with oil potential have ratified the Energy Charter Treaty, which protects foreign investments – although structures to guarantee implementation are still being put in place.

According to analysis in the recent IEA study, *Caspian Oil and Gas*, both high and low cases are presented of oil production, domestic consumption and export potential for Kazakhstan, Azerbaijan, Turkmenistan and Uzbekistan over the period 1990–2020.[5] The high scenario assumes implementation of present projects without delay, to be followed by additional development projects. It also assumes a high GDP growth scenario coupled with higher domestic demand. Tables 6.3 and 6.4 present estimates of future oil production, domestic consumption and net exports for the four important countries as well as for total Caspian oil.

In the high case, total Caspian production increases from 79 mt in year 2000, to 194 mt in 2010 and 308 mt in 2020. Net exports are projected to increase from 33.1 mt in 2000 to 116.5 (about 2.3 mbpd) in 2010 to 177.9 mt (about 3.6 mbpd) in 2020. In the low case scenario production is 69 mt in 2000, increasing to 138 mt in 2010 and 241 mt in 2020. Exports are correspondingly less: 28.6 mt (or about 0.5 mbpd) in 2000, 75.1 mt (or about 1.5 mbpd) in 2010, and 148.2 mt (or about

TABLE 6.3
Oil Production, Domestic Consumption and Net Export (high case)
(Million Tons)

	1990	1995	2000	2005	2010	2020
Kazakhstan						
Production	25.2	20.5	45.0	70.0	100.0	160.0
Consumption	27.2	10.4	20.0	33.7	45.5	84.2
Net exports	-2.0	10.1	25.0	36.3	54.5	75.8
Azerbaijan						
Production	12.3	9.2	14.0	30.0	70.0	120.0
Consumption	8.6	7.0	10.2	13.0	15.0	25.9
Net exports	3.7	2.2	3.8	17.0	55.0	94.1
Turkmenistan						
Production	3.4	3.5	10.0	11.0	12.0	14.0
Consumption	4.8	5.7	7.0	7.0	7.0	8.0
Net exports	-1.4	-2.2	3.0	4.0	5.0	6.0
Uzbekistan						
Production	2.8	7.6	10.0	11.0	12.0	14.0
Consumption	10.2	8.6	8.7	9.0	10.0	12.0
Net exports	-7.4	-1.0	1.3	2.0	2.0	2.0
Total						
Production	43.7	40.8	79.0	122.0	194.0	308.0
Consumption	50.9	31.7	45.9	65.6	77.5	130.1
Net exports	-7.1	9.1	33.1	56.4	116.5	177.9

Source: IEA, *Caspian Oil and Gas*, 1998, 51.

3 mbpd) in 2020. Because of uncertainties in the timing of large projects yet to be implemented in Azerbaijan and Kazakhstan, the largest gap between the high and low scenarios for oil exports is in 2010.

In developing the above data, the IEA used a "bottom-up" approach of the main energy-using activities in the country economies concerned to develop energy consumption forecasts. There are important differences among the countries. The Kazak economy is the largest and also the most diversified. It has the largest domestic oil consumption. In Azerbaijan, the energy sector has the largest share in GDP, about 15 percent.

TABLE 6.4
Oil Production, Domestic Consumption and Net Exports (low case) (Million Tons)

	1990	1995	2000	2005	2010	2020
Kazakhstan						
Production	25.2	20.5	40.0	55.0	75.0	130.0
Consumption	27.2	10.4	15.6	24.4	31.6	51.9
Net exports	-2.0	10.1	24.4	30.6	43.4	78.1
Azerbaijan						
Production	12.3	9.2	14.0	25.0	45.0	90.0
Consumption	8.6	7.0	10.2	12.9	14.8	21.9
Net exports	3.7	2.2	3.8	12.1	30.2	68.1
Turkmenistan						
Production	3.4	3.5	6.0	6.5	7.0	8.0
Consumption	4.8	5.7	6.0	6.5	7.0	8.0
Net exports	-1.4	-2.2	0.0	0.0	0.0	0.0
Uzbekistan						
Production	2.8	7.6	9.0	10.0	11.0	13.0
Consumption	10.2	8.6	8.7	9.0	9.5	11.0
Net exports	-7.4	-1.0	0.3	1.0	1.5	2.0
Total						
Production	43.7	40.8	69.0	96.5	138.0	241.0
Consumption	50.9	31.7	40.4	52.8	62.9	92.8
Net exports	-7.1	9.1	28.6	43.7	75.1	148.2

Source: IEA, *Caspian Oil and Gas*, 1998, 51.

Its economy is expected to receive the greatest boost from the oil and gas sector. The IEA notes:

> In potential major oil and gas exporting countries such as these, production will probably depend more on finance and export opportunities than on the full success of internal economic reforms. However, the latter can speed up the process considerably by attracting additional domestic and foreign funds and promoting the timely installation of infrastructure necessary for oil and gas development. While the higher GDP growth generated by intensified reforms will entail more domestic energy use, the incremental amounts of consumption are assumed to be less than the additional energy exports made possible through increased production.[6]

The EIA's 1998 *International Energy Outlook* includes a chart from Petroconsultants of Houston that is even more optimistic (probably too optimistic). It estimates Caspian production in 2010 at 4.7 mbpd and exports around 3 mbpd while production in 2020 is estimated around 6.9 mbpd with exports at around 6 mbpd. Projections for Caspian production by Dr Matthew Sagers of PlanEcon are lower but closer to the forecast by Petroconsultants. He projects Caspian oil production at about 4 mbpd in 2010 and 6.4 mbpd in 2020. He concludes that the Caspian area is on track to become another North Sea.[7]

How do these projected export figures for Caspian oil compare with exports of other OPEC or non-OPEC producers? According to the EIA of the US DOE, in 1995 the North Sea exported 4.6 mbpd, the former Soviet Union (FSU) exported 2.6 mbpd, and other non-OPEC countries exported 2.6 mbpd. OPEC exports from the Arab Gulf, on the other hand, were 15.4 mbpd, from North Africa 2.3 mbpd, from West Africa 2.1 mbpd and from South America 2.6 mbpd (see Table 6.5).[8]

In 2020, EIA projections for exported oil are 4.1 mbpd from the North Sea, from the FSU (including Caspian) 5.6 mbpd and from other non-OPEC 3.8 mbpd. Among OPEC members, Arab Gulf countries are expected to export 41.8 mbpd, North Africa 2.7 mbpd, West Africa 2.3 mbpd and South America 4.3 mbpd.

TABLE 6.5
World Petroleum Trade
(Million Barrels per Day)

1995

Country/Region	Exports
North Sea	4.6
FSU	2.6
Other non-OPEC	2.6
OPEC:	
Arab Gulf	15.4
North Africa	2.3
West Africa	2.1
South America	2.6
Caspian (IEA 2010 high case)	2.3

2020 Forecast

Country/Region	Exports
North Sea	4.1
FSU (including Caspian)	5.6
Other non-OPEC	3.8
OPEC:	
Arab Gulf	41.8
North Africa	2.7
West Africa	2.3
South America	4.3
Caspian (IEA 2020 high case)	3.6

Source: Adapted from EIA, *International Energy Outlook*, 1998, 36 and IEA, *Caspian Oil and Gas*, 51.

Comparing the moderate IEA numbers for future Caspian exports, in 2010, exports of 2.3 mbpd resemble the exports of OPEC-North Africa today and are less than the exports from the North Sea. In 2020, taking the IEA high-case number of about 3.56 mbpd, Caspian exports would be slightly less than projected North Sea exports (which by then are declining) as estimated by the EIA in that year (4.1 mbpd). Venezuela would be exporting more. In no case would Caspian exports, either in 2010 or in 2020, measure up to the very large exports from the Arab Gulf.

Implications for OPEC and the World Oil Market

Turning again to the EIA forecast which incorporates the most recent figures, the oil demand is expected to grow by an average of 2 percent per annum in its reference case, which will lead world oil consumption to increase from its present level of about 75 mbpd to more than 115 mbpd by 2020. (Oil demand grows faster in developing countries, slower in industrialized countries.) This is based on an estimated gradual increase in oil prices to US$23 per barrel in 2020 (1996 dollars). Transportation is assumed to be the major sector driving increased oil demand. Fairly rapid recovery is presumed for the economies of East and Southeast Asia. The EIA also presents high and low world oil price cases.[9]

In the reference case, the EIA anticipates that the increase in world oil supply of about 40 mbpd by 2020 will come one-fourth from non-OPEC countries and three-fourths from OPEC countries as indicated in Tables 6.6 and 6.7 and Figure 6.1. It is assumed that OPEC countries (especially Arab Gulf OPEC countries with low production and capital investment costs) will make the necessary investments to expand capacity either by themselves, or by attracting outside investment capital. Other assumptions include:

- expansion of OPEC production outside the Arab Gulf in offshore Nigeria, Algeria, Indonesia and Venezuela
- continued expansion of non-OPEC supply from proven reserves, including the North Sea where production is estimated to peak at 7.6 mbpd in 2003 and then begin a gradual decline

- considerable expansion of production in West Africa and in South America
- falling US output will be offset by production increases in Canada and Mexico
- oil production increases in the FSU to 9.5 mbpd by 2005 and exceeds 13.1 mbpd by 2020, of which just under 5.7 mbpd is exported (mostly from the Caspian Basin)
- continued improvements in oil production technology and no resource constraints in the forecast period.

TABLE 6.6

Non-OPEC Oil Production, 1990–2020
(Million Barrels per Day)

Year	Reference Case	High Oil Price	Low Oil Price
History			
1990	42.2	–	–
1996	43.5	–	–
Projections			
2000	47.3	47.9	46.6
2005	52.0	53.0	50.2
2010	55.0	56.7	52.6
2015	55.2	57.6	52.5
2020	55.3	57.7	52.4

Note: Includes the production of crude oil, natural gas plant liquids, refinery gain and other liquid fuels.

Sources: History: EIA, *International Petroleum Statistics Report,* DOE/EIA-0520 (97/12) (Washington, DC, December 1997), Table 1.4.

Projections: EIA, *World Energy Projection System* (1998). Reprinted from EIA, IEO 1998, 32.

TABLE 6.7
OPEC Oil Production, 1990–2020
(Million Barrels per Day)

Year	Reference Case	High Oil Price	Low Oil Price
History			
1990	24.5	–	–
1996	28.3	–	–
Projections			
2000	29.9	28.2	32.1
2005	34.3	30.9	39.9
2010	40.6	35.3	48.8
2015	49.9	42.6	60.2
2020	60.5	52.5	72.9

Note: Includes the production of crude oil, natural gas plant liquids, refinery gain and other liquid fuels.
Sources: History: EIA, *International Petroleum Statistics Report,* DOE/EIA-0520 (97/12) (Washington, DC, December 1997), Table 1.4. **Projections:** EIA, World Energy Projection System (1998). Reprinted from EIA, IEO 1998, 31.

FIGURE 6.1
OPEC Oil Production in Three Oil Price Cases, 1970–2020

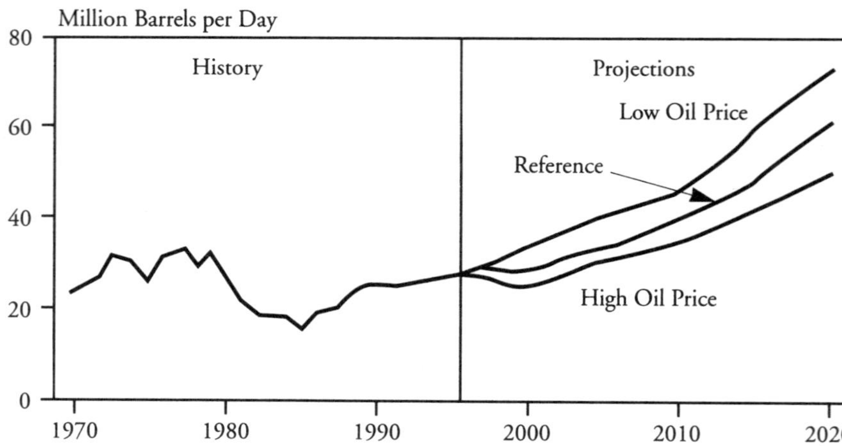

Sources: 1970–1979: Energy Information Administration (EIA), Office of Energy Markets and End Use, International Statistics Database. **1980–1996:** EIA, *International Energy Annual 1996*, DOE/EIA-0219(96) (Washington, DC, February 1998). **Projections:** EIA, *World Energy Projection System* (1998).

The EIA also presents a fourth high non-OPEC supply case in which oil prices are the same as in the reference case, but it is assumed that some undiscovered oil in non-OPEC nations will be developed and produced before 2020. In this case, the annual growth rate for non-OPEC production is projected at 1.4 percent, compared with a 1 percent growth rate in the reference case. In the high case, non-OPEC production would reach 60.5 mbpd, compared with 55.3 mbpd in the reference case (see Figure 6.2). In the reference case, the OPEC share of world production starts to exceed that of non-OPEC in 2018 (the cross-over point in Figure 6.2), while in the high non-OPEC supply case, OPEC's attaining majority market share is postponed beyond the forecast period.[10]

FIGURE 6.2
OPEC and Non-OPEC Oil Production in Two Cases, 1990–2020

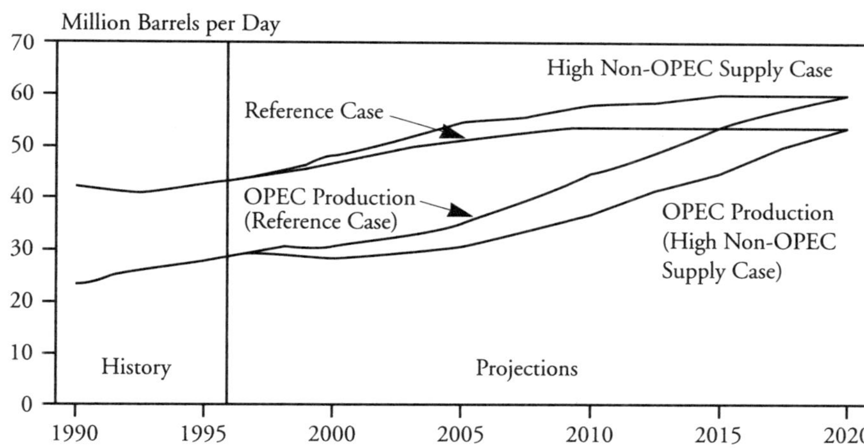

Sources: History: Energy Information Administration (EIA), *International Energy Annual 1996*, DOE/EIA-0219(96) (Washington, DC, February 1998). **Projections:** EIA, *World Energy Projection System* (1998).

My own judgment is that the EIA is using overly optimistic estimates of Caspian Sea oil exports in 2020 (based on data from Petroconsultants mentioned earlier). But the prognosis is high for continuing strong non-OPEC production from other countries and regions in view of recent and continuing advances in production technology and efficiency. Therefore, there may be a smaller gap between the reference case lines in

Figure 6.2 and the high non-OPEC supply cases, leading to a cross-over around 2020. (See also Figure 6.3)

FIGURE 6.3
Non-OPEC Oil Production by Region in Two Cases, 2020
Million Barrels per Day

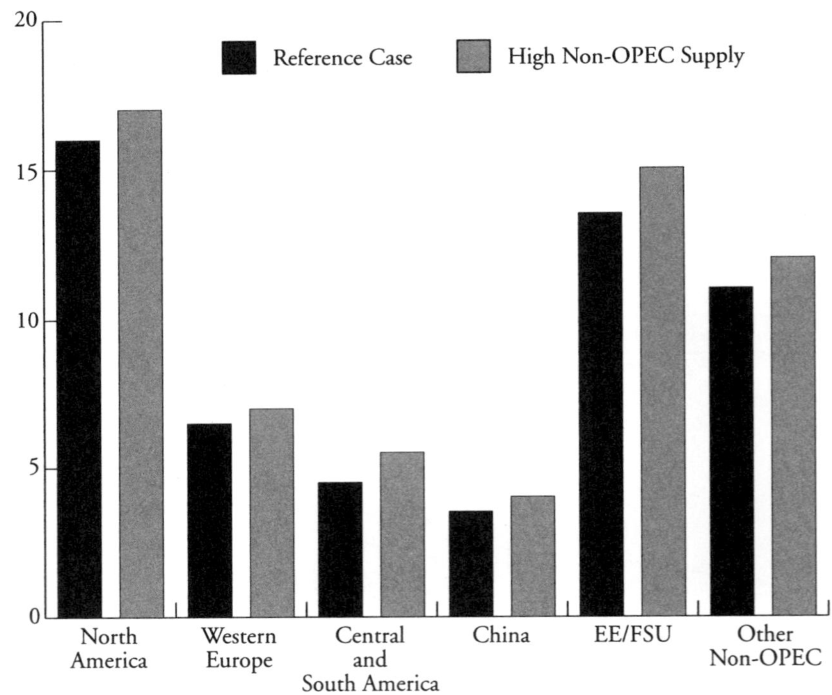

Source: Energy Information Administration, *World Energy Projection System* (1998).

What are the likely markets for Caspian oil? Certainly some of the oil will be consumed by the Caspian republics themselves and by other neighboring countries in the Black Sea region. Considering proximity and transportation costs, and assuming the solution of problems of creating pipelines, the logical markets for the bulk of oil exports are in the Mediterranean area and in Western Europe, although some oil may also prove attractive to refiners in southern Russia. As Sarah Emerson explains further in her chapter, the quality of Caspian oil also argues for

European markets. Crude oil from Azerbaijan is light and sweet and of high quality. Crude oil from Kazakhstan is light while high in sulphur and mercaptans, although most of the latter are being removed before export. Thus, this oil is directly suitable to European refiners seeking to make lighter transportation fuels under increasingly stringent environmental requirements. Caspian oil is therefore likely to compete with Middle East OPEC crude and North Sea crude in European markets up to 2010, but thereafter part of it will replace gradually declining North Sea crude oil in Europe while part will continue to compete with Middle East crude. Caspian oil is, in general, higher in quality and cheaper than Siberian crude. Caspian oil could also edge Siberian crude out of European markets. The result will be that more Middle East OPEC crude oil will be available for Asian markets where it could well be needed assuming that economic recession does not last too long.

Eventually, as quantities of Caspian oil exports grow, and new pipelines are built, some Caspian oil may also go to China, East Asia or South Asia. But given the large economic and political obstacles that must be overcome before the necessary pipeline projects become reality, it is hard to predict when this might happen – certainly not before 2010.

Implications for Energy Security – the American View

Since the end of the Cold War, oil security in the American view has first revolved around defending access to the Arab Gulf oil (as the United States did in assuming leadership of the UN coalition against Iraq in the Gulf War of 1990–91). However, the United States has also favored enhancing and diversifying global energy supplies in order to reduce dependence on oil from the Arab Gulf, which is still viewed as an unstable region where the outbreak of revolutions or wars could again interrupt oil supplies and cause future price shocks.

US policy as quoted below, with respect to development of Caspian energy resources is said to have four objectives:

- strengthening the sovereignty, independence and prosperity of the Caspian newly-independent states (NIS)
- enhancing commercial opportunities for US and other companies

- bolstering the energy security of the US and our allies and ensuring the free flow of energy from the Caspian region to the world market place and
- mitigating regional conflicts by building economic linkages between the new states of the region.

In pursuing these objectives, the US supports:

> the establishment of an east–west energy transit corridor, comprised of a network of multiple pipelines that will bring Caspian energy resources to world markets while bypassing the potential choke-point of Iran. This network will include a Baku–Ceyhan main export pipeline for oil, the Caspian Pipeline Consortium (CPC) project, the new early oil pipelines from Baku to Supsa and Novorissisk, and a trans-Caspian gas pipeline stretching from Turkmenistan to Turkey.[11]

In an era when energy policy has reverted more and more to the private sector and free market, it is not completely clear how the US government can steer pipeline decisions which in the end must meet commercial criteria. But it is trying hard. While eschewing direct financial assistance, the Clinton administration has instructed its finance and investment agencies (the Trade and Development Agency, the Overseas Private Investment Corporation, (OPIC) and US Export–Import Bank (EXIMBANK)) to coordinate their efforts to promote Caspian energy investments. And it has persuaded the five Caspian presidents to sign the Ankara Declaration declaring their support for Baku–Ceyhan, the CPC, and the trans-Caspian gas pipelines. In October 1998, Undersecretary of State Stuart Eizenstat announced a new US Trade and Development Agency (TDA) grant of US$823,000 to the Turkish pipeline consortium BOTAŞ to support planning of the Baku–Ceyhan oil pipeline and the trans-Caspian gas pipeline. Total cost estimates of the pipeline range from US$2.8 to US$4 billion.[12]

A second inter-governmental accord favoring construction of the Baku-Ceyhan pipeline was signed in Istanbul in November 1999 in the presence of US President Bill Clinton, under which Turkey guarantees to pay any costs of the Turkish segment in excess of $1.4 billion.[13] This guarantee may facilitate efforts to obtain private financing.

In keeping with its hostility toward Iran until the latter government changes policies that support terrorism and nuclear proliferation and desists from undermining the Middle East peace process, and in carrying out the Iran–Libya Sanctions Act, the Clinton administration is steadfastly opposed to major investments or pipelines routes going through Iran.

There are gaps in the new US policy to support Caspian energy development and pipelines, since the Caspian region itself is politically unstable and replete with ethnic and border disputes and open to terrorist attacks against pipelines and other oil facilities. There is a need to construct security arrangements to guard against these possibilities, but so far this has not been done.

Conclusion

It is important to keep Caspian oil development in perspective. Estimates for Caspian Sea oil have been reviewed which indicate proven reserves in the range of 15–40 billion barrels and potential reserves which could raise the total to 200 billion barrels or more. Proven reserves place the Caspian in the range of the North Sea today, though clearly the potential could be much greater. However, the potential in no way compares to the proven reserves of the Arab Gulf which are well over 600 billion barrels.

According to the moderate estimates of the IEA, which has done careful analysis of production and export potential in the Caspian region, and assuming that projects go forward without too much delay caused by low oil prices, Caspian production could reach 194 mt in 2010, of which 2.3 mbpd could be exported. These numbers could rise to 308 mt of production in 2020, of which 3.6 mbpd could be exported. These estimates place the Caspian export potential in 2010 below the level of the North Sea which exported 4.6 mbpd in 1995, but the Caspian could approach the declining North Sea export estimates of 4.1 mbpd in 2020 (using EIA estimates for the North Sea). In 2010, Caspian exports are more likely to be equivalent to North African oil exports today. Caspian exports do not approach the range of Arab Gulf oil exports, which were 15.4 mbpd in 1995 and could increase to 41.8 mbpd in 2020 according to the EIA.

The development of Caspian oil is complicated, of course, by difficult economic and political questions related to pipeline construction. But assuming that pipeline projects go forward, Caspian oil will add non-OPEC oil to the world oil market and postpone the time when

OPEC supply once again surpasses non-OPEC supply (projected to be around 2020 give or take a few years). Before that time, since most Caspian oil is light, it will compete with Arab Gulf oil in European markets, which may reinforce the tendency of more Gulf oil going to Asian markets.

US government political support may assist Caspian oil development in some ways. But in the end commercial criteria and the level of oil prices will be more important in deciding where and when pipelines will be built. Another important factor will be whether Saudi Arabia and other countries in the Arab Gulf decide to change their policies and invite international oil companies back to assist with the expansion of their upstream production. If that happened, presumably it would reduce the amount of international interest in Caspian oil development where production costs are higher than in the Gulf.

7

Price vs. Market Share for the Arab Gulf Oil Producers: Do Caspian Oil Reserves Tilt the Balance?

Amy Myers Jaffe

Introduction

The major oil producers of the Arab Gulf sit on approximately 45 percent of the world's proven oil reserves and supply 20–25 percent of current world oil demand, making the Gulf the most important oil-exporting region in the world. The Arab Gulf producers also hold over 80 percent of the current cushion of world spare productive capacity which stood at three to five million barrels per day (b/d) in 1999.

This huge capacity – both present and potential – gives the oil producers of the Arab Gulf region a unique status and opportunity but also confronts them with special challenges. This group of oil producers has sufficient market power to influence oil prices and to hold them above marginal costs. Morris Adelman argues correctly that oil prices have remained substantially above marginal costs of US$5 a barrel for several decades.[1] But with that ability comes the Pandora's box of choice: Is it wiser for the Gulf Arab producers to hold back productive capacity to promote higher world oil prices, or should the producers seek expanding market share at the expense of overall price levels? The vacillation between these two strategies has characterized the volatile nature of global oil price movements over the past two decades.

It has been argued that the huge energy resource potential of the Caspian Basin region represents a major challenge to the supremacy of the Arab Gulf as a pivotal supplier of energy to world markets and calls into question the wisdom of Arab Gulf production cutbacks designed to boost oil prices. While higher oil prices will undoubtedly encourage investment in high-cost regions like the Caspian Basin, price is not the only factor influencing the speedy development of Caspian oil resources. Rather, a host of complicated economic, logistical and geopolitical obstacles block the landlocked region's ability to become a major oil-producing province of the magnitude of the Arab Gulf or even the North Sea or Latin America.

Incremental production from Central Asia and the Caucasus can at the margin contribute to a weakening of oil price levels. But it is unlikely to be a major threat to the market share and market power of the Arab Gulf producers. Caspian oil production is unlikely to represent more than 3 to 4 percent of world oil demand by 2010. Therefore, the Caspian Basin potential alone does not justify Arab production policies that would drive down the price of oil simply to discourage Caspian oil development.

However, other more complex economic factors dictate that Arab Gulf oil producers may not be able to sustain high oil prices over the longer term, even if large market share sacrifices are made. Regardless of the reserves of the Caspian Basin, Arab Gulf producers can benefit from market share orientation. Lower oil prices may stimulate both incremental oil use as well as economic expansion that in turn will create additional demand for energy. However, oil producers from the Arab Gulf may not garner the full benefit of price-induced oil demand increases if major consuming countries react to lower energy costs by augmenting energy taxes.

The Limits of Monopoly Power in the Gulf

In considering the experiences of oil markets in the 1970s and 1980s, it is useful to think about the theory of monopoly power. Theoretically, the extent to which a monopolist can fix a commodity price above marginal cost is inversely related to the elasticity of demand for that commodity. In oil markets where – besides the Arab Gulf producers – there exists a competitive fringe of competing suppliers who produce and sell oil at prices related to marginal costs, the elasticity of demand facing the Arab Gulf suppliers is a function of global elasticity of demand and the elasticity of supply of the competitive fringe.[2]

In the short term (a year or less), the elasticity of demand or fringe supply is relatively small, giving the Gulf Arab producers some leeway to boost prices temporarily. But such a boost will be hard to sustain over the longer term, as a decline in consumer demand in combination with an increase in production from the competitive fringe suppliers can be expected. These reactions will limit the Arab Gulf producers' ability to raise oil prices in the long run without encountering a painful loss of market share.[3]

As Dermot Gately demonstrates, the OPEC discounted revenue stream is highly sensitive to uncertain and changeable factors such as price elasticities for oil demand, non-OPEC supplies (i.e. oil production that comes from countries outside the ranks of OPEC) and world income growth.[4] The experience of such uncertainty is made visible by viewing the variety of oil market modeling exercises and their sensitivity to scenario alterations.[5] Gately also shows that OPEC's discounted revenue stream was dramatically lessened by trying to increase prices during a period of slower than expected world income growth and higher than expected rising non-OPEC supply. He advocates that OPEC should have pursued a more modest increase in market share during periods of market strength since the 1970s in line with world income growth.[6]

Historical experience and economic modeling has shown that, as Arab Gulf producers move to increase revenues by reducing output, rising prices will encourage consumers to invest in energy-efficient goods and may even contribute to a slowdown in economic growth, as energy inputs become more expensive. Rising oil prices also encourage investment in new productive capacity. Companies not only have higher cash flow to make incremental investments but marginal projects are also rendered profitable. Some of these structural changes will remain even after prices begin to temper in reaction, reducing the long-run benefit.[7]

The above scenario aptly describes the experience of the 1970s oil shocks that resulted in a prolonged curtailment of the growth in the use of oil into the 1980s and in a huge boost in supplies from the competitive fringe, often referred to as non-OPEC supplies. Much has been written about OPEC's declining influence on world oil prices as non-OPEC supply has captured an increasing share of world markets. The price shock beginning in 1979 brought an impressive response from non-OPEC producers that eventually reduced OPEC's market share – from 49 percent in 1979 to 36 percent in 1995.[8]

As Arab Gulf leaders consider the resource potential of large emerging, high-cost regions like the Caspian Basin, the wisdom and attractiveness of holding back production to boost prices has been questioned. The existence of these plentiful, exploitable oil reserves outside the Arab Gulf, combined with Iraq's eventual full return to international markets, means that a high-price strategy may not be sustainable for Gulf producers. Given the dimmer outlook for the Asian economies and other emerging markets, such a strategy would require the shut-in of millions of barrels a day of existing capacity while curbing demand growth and creating incentives for the development of competing resources. Moreover, it is unclear that large-scale shut-ins of capacity would increase overall revenues even if prices were to rise for a period of time.

Forecasts for 2005–10

Non-OPEC production has expanded by 1 to 1.5 percent per annum on average since 1988. This has been accomplished during a period when prices were either stable or declining, with the brief exception of increases during the 1990–91 Gulf crisis. A combination of technological advances and discoveries of new basins in South America, in deep water and elsewhere contributed to the gains. Should this trend continue, non-OPEC production could reach 54 mbpd by 2005 and 58 mbpd by 2010, roughly 64–68 percent of anticipated global oil demand of between 80 to 85 mbpd in 2005 and 61–69 percent of forecasted global oil demand of between 84 to 94 mbpd by 2010, assuming continuation of average historical price trends. Such a forecast would imply that OPEC would need to hold back significant volumes of anticipated productive capacity for 2010 to balance supply with demand in 2005 and 2010 and promote moderate oil prices in the high teens. Given expected OPEC capacity gains for countries such as Venezuela, Iraq, Iran and Libya, among others, OPEC might need to shut-in between 8 mbpd to 15 mbpd by 2005–10 just to keep prices above current levels.[9] By contrast, OPEC shut in 2 to 3 mbpd of productive capacity in 1998.

Ironically, Kuwait, the UAE, Qatar, Oman and even Iraq are pursuing foreign private investment to raise output capacity, despite expectations of sustained low oil prices. Saudi Arabia continues to guard its market share relative to other producers and refuses to return to a "swing" producer role to defend prices. One rationale for a market share-oriented approach in the Gulf is that lower oil prices will discourage development of competing high-cost reserves such as those found in the Caspian Basin.

This market share strategy makes sense but may not be able to wrest back as much market share as the Arab producers would like. In the very short term, low prices will neither force many oil field closures, nor cancellation of major investments. Environmental and other considerations can also render it more costly to shut-in high-cost oil production wells than to operate at a loss for considerable periods of time. High-cost oil producers, burdened with short-term expenses, will likely wait to react to falling prices until it is clear that the price decline will be substantial and sustained. Some producers are hedged in forward, futures and derivatives markets and therefore will be immune to oil price swings for a period of time. This fact has shorn up US lower 48-state production (US oil production excluding Alaska and Hawaii) despite the sharp drop in oil prices in 1998. In the first 11 months of 1998, US production declined by only 60,000 b/d to 6.3 mbpd from levels of a year ago.[10] But North Sea production gains are expected to slow to about 300,000 b/d in 1999,[11] and delays are expected in some deepwater African oil field developments.

Additionally, consumers will also not react immediately to a dip in oil prices. For example, declining gasoline prices will not stimulate immediate purchases of larger cars or more driving in the initial stages of a price change. Businesses may not expand to take advantage of lower energy costs, fearing that OPEC will eventually intervene and lower prices will not last. The same can be said for planned investments in expensive oil field projects. Oil company planners must believe that oil prices will remain low over the period of the project development – which could take 20 to 30 years. Moreover, forward financial derivative markets allow investors to lock-in long-range oil prices that may protect the project from future losses.

Over the longer term, improvements in oil drilling, finding and production technologies can be expected to protect the output from high-cost production regions as seen over the past decade or so in the

North Sea and US Gulf coast deep water. Alarmists have argued that "most of the apparent improvement in recovery factors is an artifact of reporting" and that global oil production will decline precipitously within the next ten years as major basins discovered in the 1970s move into massive decline.[12] But history and predictions of emerging technologies tend to show otherwise. Michael Lynch of the Massachusetts Institute of Technology (MIT) shows the systematic bias of predictions of production outlook and their underestimation of the impact of technological change.[13] Improvements in cost reduction and technological breakthroughs have considerably lowered the profit threshold for previously expensive programs such as deepwater drilling and oil shale recovery. Moreover, science is bringing progress to understanding the migration patterns of oil under the ground, and breakthroughs could yield substantial new ideas about reserves in mature regions like the United States.[14]

Technology and improvements in cost reduction have significantly narrowed the gap in production costs between the Arab Gulf oil producers and Western provinces such as the US Gulf of Mexico and the North Sea. Lack of ongoing infrastructure investment and inefficient state monopoly practices have increased Arab Gulf production costs to between US$2 and 3 a barrel, from about US$0.10–0.50 a barrel in the 1970s, according to studies by the Centre for Global Energy Studies (CGES) and Energy Intelligence Group. US and UK fields, by contrast, have dropped from an average of US$12–15 a barrel to US$4–7 a barrel over the last decade or so, making it that much more expensive for OPEC to force shut-in of non-OPEC production through large market price decreases.

Government policy also has a big role to play in protecting high-cost production. Taxation and fiscal regimes play a major role in determining incremental production rates as Britain's tax cuts in the 1980s and Venezuela's oil sector liberalization of the early 1990s clearly demonstrate.[15]

Eventually, nonetheless, there is some evidence that in the longer run, lower oil prices will stimulate increased consumer demand for oil. On a worldwide basis, many analysts believe global price elasticity of demand is unitary or less, meaning that the revenues of the world's oil producers would not be enhanced through consumption additions alone as prices fall. But, to the extent that the Arab Gulf producers attract some of the limited foreign exploration investment dollars for their own fields, thereby diverting significant investment from oil fields outside their domain, Gulf market share could be enhanced further. Moreover, if lower oil prices can stimulate economic expansion of the economies of oil-consuming countries, oil producers – in particular, those large producers of the Arab Gulf – can benefit.

The extent to which lower oil prices contribute to economic growth remains under debate. Several economic studies have questioned whether there are significant correlations between oil price decreases and GNP increases in the United States, Canada, Japan, the United Kingdom and West Germany.[16] Recent studies suggest that, at least in the short term, a large drop in oil prices may also lead to an initial reduction in GDP growth in the US economy.[17] Even over the longer term, some indicators such as employment may also not react positively to lower energy costs. In a recent paper on job creation and destruction, Steven J. Davis and John Haltiwanger found an asymmetric response to oil price shocks in the employment sector. They observed that while job destruction was found to be severe in the case of a major sudden increase in oil prices, a commensurate scale of job creation did not emerge when prices fell back to a similar extent.[18]

In 1988, as European and US oil use was growing and international oil prices were dropping considerably, individual fuel and gasoline consumers saw virtually no passthrough of low prices at the retail level. Instead, oil taxes and oil refiner corporate profits gained much of the benefit of the sudden decline in crude oil prices. Final consumer prices hardly shifted in high-tax countries such as those in Europe and the Far East and fell only slightly in the United States, despite a 35–40 percent fall in spot crude oil markets. Oil taxes in Japan were actually increased in 1988, thwarting any passthrough of lower prices to consumers. In 1986, only half of the benefit of cheaper crude prices was passed on to Western consumers.[19]

The possibility that additional energy taxes could be imposed in consuming countries could hinder the Gulf Arab producers' ability to garner enough sustainable additional market share for themselves at a lower oil price level to boost revenues overall. Energy taxes as of 1996 represented anywhere from 29.79 percent (the US) to 69.38 percent of the total gasoline price to consumers in the OECD.[20] If oil producers are less willing to change output levels than consumers are willing to buy less fuel, it is likely that oil producers will bear the brunt of the cost of increased energy taxes.

Competition from the Caspian Basin?

Despite some uncertainty about the benefits of lower oil prices, it has been discussed in the Arab Gulf whether these producers can keep out the competition of the Caspian Basin and preserve their market power by pursuing high production policies that will lead to lower prices. While the benefit to the Arab Gulf of a more moderate oil price strategy as opposed to an unsustainable high-price strategy seems apparent, it remains unclear whether the Caspian Basin's promising geology in itself really provides true impetus for an all-out price war strategy.

The hydrocarbon reserves of the Caspian Basin are significant. Proven oil reserves are pegged at between 15 and 31 billion barrels while future exploration may confirm that the region holds between 60 billion and 140 billion barrels of oil.[21] On the basis of such reserve estimates, it may seem reasonable at first glance to assume that these sizeable reserves represent a threat to the market control of Arab Gulf producers. On the basis of the theoretical production volume that could be supported by Caspian geology, a low oil price strategy that would delay the competition from the region seems sensible. Caspian production costs are estimated to run in the US$5 – 6 a barrel range, and pipeline transportation capital costs and transit tariffs are likely to add another US$3 – 5 a barrel cost before exports are even seaborne. Such high costs would make Caspian investment highly vulnerable to downward oil price movements.

But, it is questionable whether Caspian investment could actually challenge the market power of the Gulf Arab producers in the coming years, regardless of the international price of oil. Over the next 10 to 15 years, Caspian Basin development may not achieve its potential, limiting its impact on the market dominance of the Arab Gulf producers. That is because a host of complicated technical, economic, logistical, geopolitical and social obstacles block the speedy development of Caspian oil reserves.

First, Caspian hydrocarbon resources are located at a great distance from the world's major energy-consuming regions. The countries of that region are landlocked. The region's producers cannot simply ship oil by tanker from domestic ports to international sea-lanes as is done from the Arab Gulf. Instead, the Central Asian and Caucasus states must rely on expensive pipelines built through neighboring countries as the chief means of transport. Many possible routes through Turkey via Georgia or Armenia, Iran, Russia, Georgia, China and Afghanistan have been considered.

However, most of the routes in question suffer from a variety of almost overwhelming security issues related to regional political uncertainties and thorny ethnic disturbances. Security arrangements remain undefined in most areas, rendering the region prone to instability and political unrest. Furthermore, the abrupt end to Moscow's rule over the region in 1992 created a power vacuum that has unleashed geopolitical competition in the region. A number of world and neighboring powers look at the region as a locale for playing out a broader contest for spheres of influence. The United States, Russia, Turkey, Iran, China, Pakistan, India, Europe, Japan, Saudi Arabia and even Israel have courted the region's leaders to improve relations and tap commercial possibilities. The rivalry among them has so far blocked the development of a major export route from the region.

Political barriers have so far inhibited a steady stream of major energy exports from Caspian countries through Russia or Iran. Moscow's bureaucratic chaos, corruption and weak administrative control have hindered routes involving sections of Russia's vast pipeline network. Moreover, certain elements inside Russia's foreign policy apparatus would like to block exports from Kazakhstan, Azerbaijan and Turkmenistan on nationalistic and economic grounds.[22] Security concerns have also plagued routes that involve the Caucasus countries and eastward pipelines through Afghanistan which suffers under the strain of civil war. Physical, environmental and political constraints on tanker transit through Turkey's Bosphorus Strait also thwart development of Black Sea routes while more expensive on-land routes from Kazakhstan to China or through the length of Turkey face steep financing challenges. Drilling constraints may mean it will be many years before enough export volumes exist to support a major export line, while political and economic problems may restrict the export of available output even if it grows for the foreseeable future.

The region is also far from major supply centers for exploratory equipment and faces a debilitating shortage of modern drilling platforms and other related supplies. Despite huge demand for equipment, there are only two assembly yards equipped for manufacturing or refurbishing offshore drilling rigs for the region: one at Astrakhan in Russia along the northern Caspian and one in Primorsk, near Baku. This has limited local supply. The constraints on infrastructure, drilling equipment and rigs are more severe in the Caspian Basin than practically anywhere else in the world. This means oil wells take considerably longer to complete, in some cases up to two years as compared to two to three months in other parts of the world.

Such logistical obstacles mean that while its energy resources may be geologically equivalent in scale to the North Sea, the region's output is unlikely to reach that potential. North Sea production has risen from roughly 2 mbpd in 1980 to 6.1 mbpd today or 8 percent of current world demand. By contrast, after 2 decades of development, Caspian oil production may top no more than half that and represent little more than 3 to 4 percent of world oil demand by 2010. Available exports to compete with Middle East oil in Europe and Asia could be as little as 2.5 to 2.8 mbpd by 2010 of which perhaps as much as 1 to 1.5 mbpd could remain within the hard to reach Black Sea region.[23]

This more conservative outlook for Caspian output suggests that Arab producers' market control may remain relatively unaffected by the existence of vast Caspian reserves in the short to medium term. Moreover, given already sunk costs of international investments there, an uncomfortably large price decrease might be required to halt development in the Caspian Basin. Therefore, while overall healthy non-OPEC growth rates might still dictate that Arab producers would benefit most from a market share strategy, the reserves of the Caspian may not be the key critical contributing factor.

Conclusion

In conclusion, the countries of the Arab Gulf must consider the limited short- to medium-term export potential of the Caspian Basin, not the region's physical geologic wealth, in determining the latter region's importance in international world oil markets. Incremental production from Central Asia and the Caucasus can at the margin contribute to a weakening of oil price levels. However, at 3 to 4 percent of world oil supply by 2010, it will not be a significant threat to the market control and market share of the Arab Gulf. The Caspian Basin alone should not be perceived as justification for the countries of the Arab Gulf to alter their oil production policies in order to discourage investment there.

Nonetheless, incremental production from Central Asia and the Caucasus can at the margin contribute to a weakening of oil price levels.

Moreover, Arab Gulf producers can benefit from low oil prices to the extent that such price levels contribute to a rise in oil use, creating an opportunity for sustainable market share expansion and giving investors extra incentive to channel exploration capital into low-cost areas such as the Arab Gulf.

8

Gulf Oil vs. the Oil of the Caspian Sea

Fadhil J. Chalabi

No oil-producing area in the world has attracted more attention and generated more controversy than the Caspian Sea. Fanciful estimates have even tried to rank Caspian oil reserves at as high a level as that of the major oil-producing countries in the Gulf. Some talk of the Caspian as if it were a new Kuwait. Other sources have gone as far as ascribing potential recoverable oil reserves of 200 billion barrels to the area. However, according to the *BP 1997 Statistical Review of World Energy*, the Caspian Sea's recoverable reserves are at present estimated at less than 17 billion barrels, or 1.5 percent of the world's total recoverable reserves. Apart from the limited size of the reserves, the area's oil is very costly to find, develop, produce and transport to world markets. Perhaps even more problematic are the very intricate and risky political and geopolitical implications of the new situation, now that four new oil-producing republics have emerged from the collapse of the former Soviet Union (FSU) with a "nationalistic" sentiment to cast off the Russian tutelage that has been imposed on them under the defunct Communist regime. The Caspian "Sea" is practically a landlocked area, and the economic and geopolitical problems arising from transporting by oil pipelines through other countries add to the risks of investments there.

With these apparent disadvantages of the Caspian Sea oil in mind, the question ineluctably arises: why the rush of so many American and other oil companies to invest in this region? Under normal market conditions, investors would naturally turn to the abundant, low-cost oil of the Gulf, rather than to these high-cost, politically hazardous areas.

Apart from the economics of investing in Caspian Sea oil (which will be discussed in detail), political motivation should not be ignored.

The declared US policy has been to encourage investing in the Caspian, and to create US interests in the geopolitically sensitive area situated near both Iran and Russia, the latter attempting to keep up its political influence in the newly independent republics after the collapse of the Soviet Union. It is also a declared policy of the United States to develop Caspian Sea oil in order to reduce dependence on Gulf oil, especially oil from both Iran and Iraq.

However, political factors aside, the rush to Caspian Sea oil was spurred on by the oil market perceptions during the period which followed the collapse of the FSU and which lasted until 1998. What further motivated this rush was the eagerness of the emerging Central Asian republics to invite foreign investors and offer them attractive conditions.

From the early 1990s until 1997, oil market perceptions were characterized by:

- The ability of the OPEC to maintain a fair degree of cohesion amongst its members which helped to stabilize oil prices at artificially high levels and for a long period. The embargo on Iraq's oil was instrumental in maintaining OPEC's supply regulatory system, because it allowed many OPEC countries, especially Saudi Arabia and Venezuela, to increase their market share substantially without damaging price stability. OPEC's capacity to sustain oil prices at high levels encouraged investment in new oil, despite the economic and political risks arising from investing in that part of the world.

- The oil technology revolution led to a spectacular reduction in the cost of finding and developing high-cost oil, so that even when the OPEC price declined from US$28 per barrel in 1985 to US$17–18 per barrel thereafter, the economics of investments in the area were still favorable. Investing company profits were enough to induce them into further investments towards expanding their operation in this high-cost area.

- Between 1994 and 1997, non-FSU world demand for oil was growing at robust rates with annual incremental demand exceeding 1.4 mbpd, a good part of which (41 percent) originated in the Asia Pacific, whose economies were growing at fabulously high rates.

- Until the mid-1990s, most OPEC countries were persistent in the policy of closing their national industries to foreign investors, a factor that pushed the latter to find alternative investment opportunities.

OPEC Policies and the Emergence of "New" High-Cost Oil

All these favorable factors and market perceptions justified economically the rush to the Caspian Sea, notwithstanding the high costs of developing the new oil and transporting it through difficult, costly and politically hazardous routes. The oil companies' rush to the Caspian was in line with the big wave of oil investments in frontier areas outside OPEC, where costs are very high and investment risks are not negligible. This happened as a result of OPEC's past policies, which, with their self-defeating and uneconomic approach to oil, encouraged investments in those areas, to its own detriment. The result was that OPEC itself, especially the Gulf producers, paid a very high price in terms of losing market shares in favor of the newcomers.

- OPEC's past price-shock policies (which took the price of oil from US$3 per barrel in mid-1973 to US$34–36 per barrel in 1981) secured a very high margin of profits for investments in "new," high-cost oil which surpassed all the incumbent economic risks. For the whole of the 1970s and for part of the 1980s, OPEC policy makers naively thought that no matter how high the price they imposed on their oil, the market demand for their oil would increase. Unfortunately for them, the excessively high prices led to, among other things, investments in upstream oil outside their territories in new production capacities, irrespective of cost differences.

- More important, perhaps, is that in its policies of keeping oil prices at artificially high levels, OPEC adopted the production quota system, which secured, for the new high-cost producers, marketing precedence over OPEC itself. The system is based on the concept of swing producer (residual supplier) which means that OPEC limits its production (the ceiling, which is the sum of all members' quotas) to the difference between world demand, on the one hand, and the oil supplies from outside OPEC including the FSU's net export, on the other hand. Buyers lift non-OPEC oil first and then turn to OPEC oil to fill the gap of their requirements. By doing so, OPEC, and especially the Gulf producers, became suppliers of the last resort with a shrinking market share.

- Until the very late 1980's, OPEC followed the system of fixed prices of crude oil under which members are not allowed to sell their crude but leave the refined product to be priced according to market forces. This meant (i) crude oil sellers from outside the Organization were free to sell according to the market price, thus enjoying a competitive advantage over OPEC and crude oil-sellers; (ii) crude oil sellers within the Organization (like Saudi Arabia) were placed under pressure to reduce their production in order to safeguard the price – a system from which other OPEC members with more sales of refined oil products (like Venezuela) benefited. While OPEC generally had to reduce its production and lose market share in order to keep prices at an artificially high level, thus carrying the burden of price protection from which non-OPEC producers benefit, Saudi Arabia and certain other member countries became the swing producers within OPEC itself, carrying an even greater burden to protect the price, and hence losing much greater market share than the other members. The beneficiaries of these policies were the new, high-cost oil producers, at the expense of the Gulf.

- To put themselves at an even greater disadvantage, OPEC policies of full state ownership (except in the case of the UAE, which retained for the national oil company 60 percent of the concession of the former Iraq Petroleum Company (IPC) Group, leaving the remaining 40 percent to the latter) benefited the new high-cost producers, mainly at the expense of the Gulf. Full state ownership of the oil industry in the Gulf led to a massive shift of capital away from the Gulf towards upstream investment in new, high-cost oil. Following wholesale nationalizations of the oil industry in OPEC (either through legislation, as in the case of Iraq, or by agreement with the oil companies as in the case of Saudi Arabia and Kuwait), international oil companies started looking for new investment opportunities outside OPEC. It is estimated, for example, that during the period 1980–95 over US$350 billions were invested in the upstream sector of the oil industry in new areas. This contributed to a great expansion of new oil development, at the expense of OPEC.

North Sea oil is a good example of how OPEC policies were instrumental in shifting investments from low-cost, oil-abundant areas in the Middle East to limited, high-cost areas. According to the BP-Amoco Annual Statistical Review, North Sea recoverable reserves stand now at about 17 billion barrels, similar to that of the Caspian Sea (1.5 percent of the world total). OPEC's policies of the 1970s and 1980s created favorable conditions for increasing upstream investments in the North Sea, which led to continuous additions of new reserves, either through the discovery of new oil fields or through upgrading already existing oil reservoirs, by means of advanced oil recovery techniques.

Heavy investment in enhancing and upgrading reserves from a limited resource base was reflected in spectacular increases in production. In 1973, the North Sea's recoverable reserves were estimated at 14.5 billion barrels, against 16.7 billion barrels now. In the meantime, production grew from zero to almost 6.3 mbpd in 1998, and the cumulative production throughout 1973–97 amounted to 28.7 billion barrels. In other words, additions of new reserves in the North Sea since 1973 were more than three times the size of reserves that had been estimated then.

Thanks to OPEC's artificially high prices, North Sea production prospered, despite its very high cost. During the 1970s and 1980s, the average combined fixed and variable cost per barrel was about US$6 per barrel, compared with less than US$1 per barrel in the Gulf. World market oil prices were so high as to justify investing in the area, despite heavy government taxation during that period – so that the per-barrel tax-paid cost incurred by the oil companies exceeded US$20 per barrel.

Yet, with an OPEC price of US$28–30 per barrel, huge profits were made for the investing oil companies in the North Sea which secured a steady cash flows for reinvesting in new expanding production capacities. Furthermore, technology reduced costs substantially by at least 40 percent which, together with host government tax-relief policies, reduced drastically the per barrel tax-paid cost. The result was that even at the lower OPEC price of US$16–18 per barrel after 1986, margins were good enough for new expansion projects.

Owing to its very high costs, North Sea oil would never have expanded without OPEC's policy of keeping artificially high prices through reducing its production in favor of the newcomers.

High-cost New Oil and the Share of Arabian Gulf Oil in the World Market

Following all these developments which favored investments in high-cost areas outside the Gulf, sweeping changes have taken place in the geography of oil and in the respective market shares. A new phenomenon presents itself, whereby oil from a limited, rather uneconomical resource base and high-cost oil from frontier areas has been replacing oil from OPEC and especially the prolific, low-cost oil of the Gulf.

World oil production outside OPEC, the US and the FSU (i.e. new oil) increased from 10.5 mbpd in 1977 to 26.6 mbpd in 1997, or an annual increase of about 850,000 barrels per day. North Sea production alone increased during that period from about 1 mbpd to 6.5 mbpd. Accordingly, "new" oil's share in meeting non-FSU world consumption increased from 20 percent to 40 percent during the same period.

This spectacular expansion of new, high-cost oil was at the expense of the Gulf. In 1997, the combined production of the five major Gulf producers (Iran, Iraq, Kuwait, UAE and Saudi Arabia) was about 1.8 mbpd lower than that of 1977, at a time when non-FSU world consumption increased by 17 mbpd.

This means that the Gulf's oil share in meeting world consumption fell from 42 percent in 1977 to 29 percent in 1997. Although OPEC, as a whole, was losing market share in favor of the "new oil," it was Gulf oil that lost most heavily from this change, caused by OPEC price policies. The share of *non-Gulf* OPEC members in world consumption fell by only 1.7 percent during these 20 years, compared with the Gulf's loss of 13 percentage points.

FIGURE 8.1
Producers' Shares of World Oil Demand (ex-FSU)

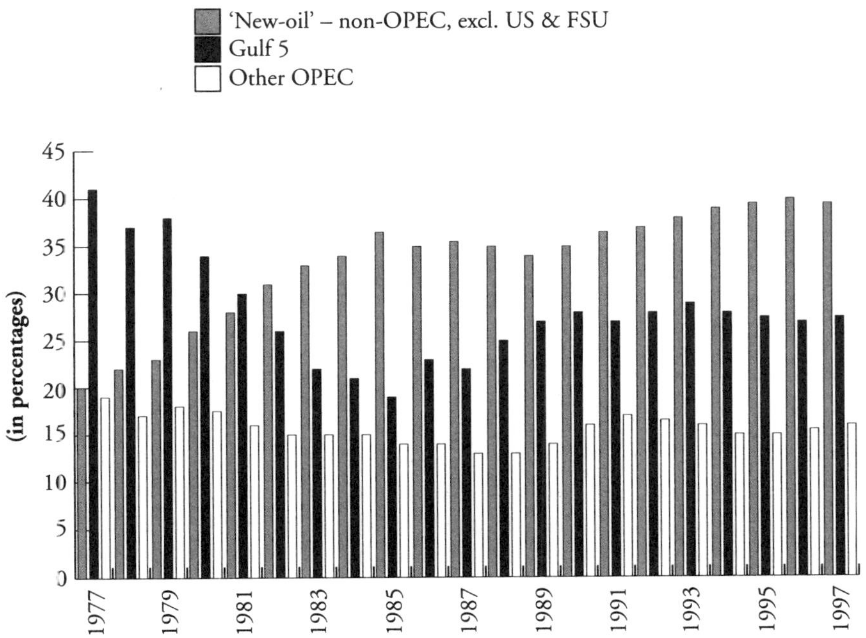

Source: CGES

Caspian Sea Oil and World Oil Prices

The future of the Caspian Sea and its impact on Gulf oil will depend crucially on oil prices and on the investment policies of the major producers of the Gulf region itself. If the present low price levels persist in the coming five to seven years (as will be discussed below), Caspian oil will have little chance of expanding. Low prices will squeeze investors' profits so much that there will be no economic incentive to invest in new projects. By contrast, if financial pressures in OPEC succeed in restoring an artificially high price of US$18 per barrel and above, Caspian Sea oil will have every chance of expanding to a similar extent as the North Sea. With the new technology, it is possible to increase recoverable oil resources from their present limited size of 17 billion b/d by three or even four times, as was the case in the North Sea. It is only with artificially high prices that the Caspian's high costs and geopolitical risks can be justified. It is high prices artificially maintained by OPEC that secured sufficient profits and so justified investing in high-cost Caspian Sea oil. Nominal fixed and variable costs per barrel in the Caspian area are on average four times those of the Gulf. Apart from the relatively prolific Tangiz oil field, nominal development in Kazakhstan could reach more than US$4.5 per barrel and average around US$3 per barrel, with a further US$3 a barrel representing operational costs.

Furthermore, Caspian sea oil could be burdened with costs from constructing new infrastructures and high-cost imports of human and technological resources. In reality, this area, which for so long was under Soviet domination, has been severely neglected by the Moscow central planners in favor of the development of Russian oil. Its infrastructure is so archaic and underdeveloped that a huge amount of capital would be needed to reconstruct it. More important perhaps are the geopolitical problems and high costs of oil transportation from the Caspian through to the world market. At present, the transportation cost to the Black Sea is more than US$4.5 per barrel plus the cost of freight to European destinations.

As far as investing in pipelines is concerned, the average long-term costs could be lower, depending, of course, on the route. The cheapest route is via Russia, an option not favored by either the US or the Central Asian republics. The political preference is to build pipelines via Turkey rather than via Iran, despite the economic logic of the latter route and the much higher costs of the former, resulting from the longer distance and the topographical nature of the region. It is generally thought that the average cost of the pipeline would be around US$3 a barrel.

In other words, the total buildup costs for new projects in the Caspian Sea, including the transportation of oil to European markets, could reach as much as US$10 a barrel. With the price at, say, US$13 a barrel, and with the share of profits being 80 percent in favor of the host government and 20 percent for the oil companies, the profit margins of investment would be much less than US$1 a barrel, which could never justify the investment economically.

What will happen to the price of oil, in other words, will be crucial in determining the size of Caspian Sea oil and its contribution to world oil supplies.

Oil Price Trends in the Medium Term

Prices will depend on the probable movements of world demand for oil and its supply. On the demand side it is expected that growth in consumption to the year 2005 will be weaker than in the last seven years or so. During the period 1990 to 1997, incremental non-FSU world demand has been about 1.4 mbpd per year. This cannot be repeated because of growing pressure on the consumption of oil resulting from many economic and political developments.

Four sources of pressure will dampen the growth of world demand in the future:

1. *Projected lower worldwide economic growth*
   The average annual growth rate of the world economy through to the year 2005 could be as low as 1.9 percent, compared with 3.3 percent over the period 1990–97. Following the Asian crisis, the deep economic recession in Russia, possibly also in Latin America and even in China, prospects for world economic growth appear much less bright than before. Most forecasts indicate that even in the United States and Europe past high economic growth rates may not be repeated in the future. (See Fig 8.2)

FIGURE 8.2
World Economic Growth Rates Excluding the Former Centrally-planned Economies

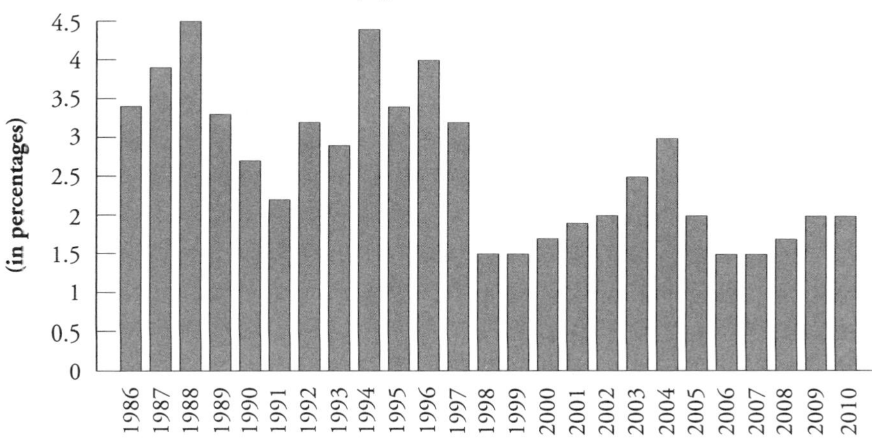

Sources: International Monetary Fund (IMF) and Centre for Global Energy Studies (CGES)

2. *Environmental pressures: the Kyoto Protocol*
   Despite serious obstacles, the target of reducing carbon-dioxide emissions by lessening oil consumption will be partially enforced. The environment has become part of the political life in the OECD countries, with an increasing power that can affect energy policies. Green movements will continue to put downward pressure on energy consumption.

3. *Higher taxation in consuming countries*
   The tax component of the composite barrel price in the OECD countries increased from 22 percent to 49 percent between 1980 and 1996. Government taxation of products in the consuming countries dampens demand growth because end-consumers pay a very high price internally, no matter how low crude oil prices may be in the world market. In Western Europe, taxes represent about two-thirds of the price of composite barrel paid by end-consumers. In certain cases, the tax component of the price of gasoline could reach 80 percent. This trend of increasing taxation may continue, and be adopted by other consuming countries.

FIGURE 8.3
Oil Taxation in the OECD

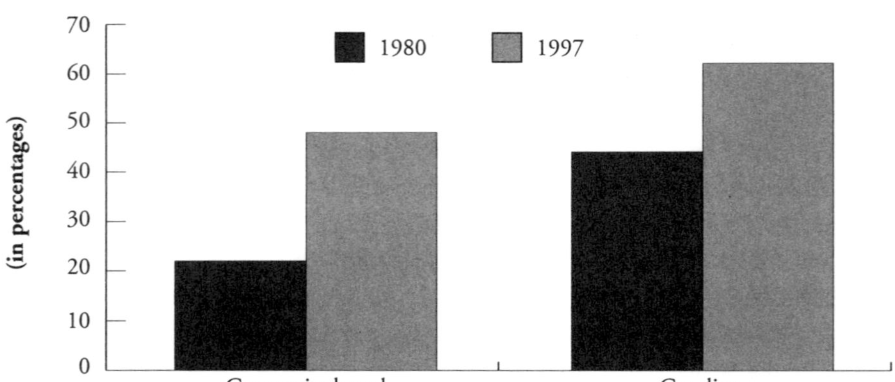

Source: CGES

4. *Higher efficiency in fuel utilization*
   Decline in oil intensity (i.e. the amount of oil needed per unit of GDP growth) may continue as a result of technological progress. This means that oil required for world economic growth, especially

in the industrial countries, will be less in the future, as the efficiency of fuel utilization will increase.

FIGURE 8.4
Oil Intensity Per Unit of GDP

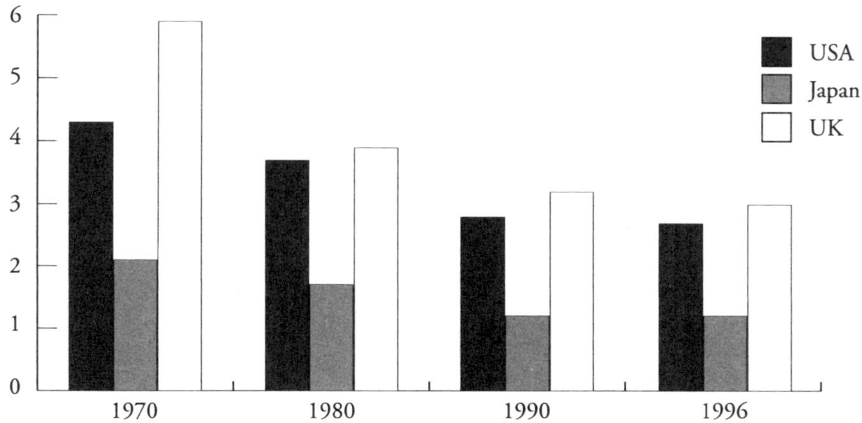

Sources: IMF, BP & CGES

On the supply side, increases in new production capacities may be realized, which could substantially outstrip growth in demand. Planned new production capacities in OPEC countries in the year 2005 may range between 11.7 mbpd and 8.6 mbpd, depending on the future of Iraq's oil and the lifting of sanctions. Other additions of some 4 mbpd new capacity may also be realized from non-OPEC areas.

FIGURE 8.5
Future Oil Supply

Projected non-OPEC supplies 1998–2005
The plans for new additions may add more than 4 mbpd by 2005.
Weak prices may cause delays or cancellations of new projects.

OPEC's expected capacity additions 1998–2005

Iran	1.6 mbpd	Saudi Arabia	0.8 mbpd
Kuwait	0.9 mbpd	Qatar	0.3 mbpd
UAE	0.9 mbpd	Libya	0.5 mbpd
Nigeria	0.6 mbpd	Algeria	0.4 mbpd
Venezuela	1.6 mbpd	Indonesia	0.0 mbpd
Iraq	4.1 mbpd (fully rehabilitated):	OPEC Total	11.7 mbpd
	1.0 mbpd (sanctions continued):	OPEC Total	8.6 mbpd

Source: CGES

Demand growth cannot outstrip this expansion in future oil supplies. It should be mentioned, however, that low oil prices may discourage part of these new capacities. If low prices persist they could trigger a reversion back to the Gulf and a new pattern of market sharing. On the one hand, low prices could lead to a severe decline in investments in non-Gulf areas, because under a low-price scenario the costs will be too high to justify investments in the new capacities; on the other hand, the cost for a new capacity in the Gulf would be so low that, despite low prices, oil companies would make good profits that would outweigh the low-price regime. This, however, will depend on two major political developments:

1. Whether the trend of opening the oil industry to foreign investors, strongly initiated in Venezuela and outside the Gulf, will reach the Middle East, especially Saudi Arabia and Kuwait. It is generally believed that sooner or later these two major oil-producing countries will open their industry to foreign investors not only because their financial situation deters their ability to provide the necessary funds for their own projects, but also because the opening up of their industry will lead to a speedy expansion of market outlets for their prolific oil.

2. What will happen to Iraqi oil? In the event of a total lifting of sanctions, Iraq, with its prolific oil fields, would in a very short time expand its oil production capacity enormously. Iraq has giant oil fields that were discovered in the 1970s but not developed because of the war with Iran and the sanctions imposed following the Gulf War. A number of agreements and contracts had been signed or drawn up with Russian, French, Italian and various oil companies to develop certain giant oil fields with a capacity target of 2.85 mbpd of additional capacity. With the rehabilitation of Iraq to its pre-Gulf War level of 3.5 mbpd this addition would push Iraq's capacity to over 6 mbpd in a matter of five years from the lifting of sanctions. By contrast, if sanctions are maintained, the picture of Iraqi oil and its contribution to the world oil supply will be radically different. The best Iraq can do with the help of the UN, is to rehabilitate the oil industry, reaching a maximum of 3.3 mbpd.

With this planned increase in OPEC supply capacities (especially with Iraq's oil expansion plans in the event of the lifting of economic sanctions) against very weak demand, oil prices are bound to be weak because the over-supply in the market will be so great that only an effective OPEC program of production cuts could restore a balance to the prices. However, OPEC's capability for supply regulation through the quota system has been eroded and the Organization has become ineffectual as a price maker. The oil abundance, together with the increasing pressure on OPEC to cut further any production, will be a major set-back to world oil prices. However, the possibility cannot be excluded that financial pressures arising from the fall in oil prices may force the major oil producers to continue with *ad hoc* arrangements outside of OPEC in order to constrain production and protect the price, or increase it for budgetary considerations.

Conclusion

Caspian Sea oil production targets of three to four mbpd by the year 2010 can be achieved only when oil prices exceed the US$17–18 per barrel range. It is from within this range that a minimum of US$2 a barrel profit margin for the oil companies can be secured and thus justify the economic and political risks arising from investments in new Caspian projects. By contrast, if the present price (US$11–13 per barrel for Brent) continues, this will deter investors from investing in new oil in the area. In this case, the Caspian Sea region would give not only a negligible profit margin for the investors, but also companies may be further discouraged because of serious geopolitical problems arising from the transportation of Caspian Sea oil through land-based pipelines. Oil companies can be encouraged to invest in the region only if profits are high enough to warrant taking the economic and political risks that come with this investment. *Oil prices will be the key factor in the expansion of Caspian Sea oil.*

9

The Relevance of Caspian Oil for the World Market

Sarah A. Emerson

Beyond Pipeline Politics

In the energy community, the term "pipeline politics" refers to the ongoing machinations of several national governments, countless energy companies and thousands of individuals all bent on releasing the enormous oil resources of the Caspian Basin to the world economy. While the twin issues of route and cost are essential to any discussion of Caspian oil, the subsequent issue of the oil's impact on the global market is equally important. Will Caspian oil redraw the global energy map? How will it influence prices, trade flows and energy security? What are the implications for Arab Gulf producers? The answers to these questions depend on one's outlook on the global oil market over the next 10 or 20 years.[1]

The Long-term Outlook – Return to Business as Usual

The Asian economic crisis cut global crude oil demand by two-thirds in 1998. Even in the countries hardest hit, however, it is unlikely that oil demand can contract by a comparable amount in 1999. Indeed, it is likely that Asian oil demand will begin a modest recovery by 2000. Then, global oil demand could return to a 2 percent per annum growth rate by early in the new century. As a result, the starting assumptions of the long-term oil outlook presented here describe a return to steady growth in world oil demand. This precludes a global recession or overt government policies aimed at conserving oil or substituting other fuels for oil. On the supply side, the forecast assumes steady, but not extraordinary, growth in non-OPEC output, especially in Latin America, Africa and the former Soviet Union (FSU). By the middle of the next decade, development of non-OPEC supplies in some remote regions could accelerate in response to somewhat higher nominal prices between now and then.

This is especially important in factoring in China, which is likely to present the most uncertainty in terms of its contribution to the Asia–Pacific region's oil requirements. This forecast assumes fairly steady growth in Chinese demand after 2000 within the constraints of the conventional "stop–go" economic approach of the Chinese government. More importantly, it assumes a neutral to optimistic assessment of the development of Chinese oil reserves after 2004. During the next 10 years, China will be forced to improve the legal and financial environment for foreign investment in order to stem its growing oil import bill.

Another important baseline assumption in a "business as usual" outlook is the absence of major political or military surprises that would markedly inhibit demand or supply. Inherent in this assumption is the peaceful lifting of the embargo on Iraqi oil. The basic assumptions are summarized in Table 9.1.

TABLE 9.1
Long-Term Crude Oil Supply and Demand: Return to Business as Usual

	Yes	No
Global demand growth averaging 2 percent per annum after 2000	X	
Recovery of Asian economies (incl. China) within 2 years	X	
Return of Iraq after 2000	X	
War or military conflict that affects the flow of oil and world prices		X
Environmental legislation that markedly reduces oil demand		X
Acceleration in non-OPEC output (incl. China) after 2004	X	

Global Crude Oil Demand

The demand side of this baseline forecast is straightforward. Global oil demand growth of roughly 2 percent translates into approximately a 1.7 mbpd increase in demand per year. Most of this demand is in the form of refinery throughput and may require the construction of refining capacity or substantial de-bottlenecking (re-commissioning where appropriate) in some regions. Crude oil demand also includes the direct burn of crude oil for power generation, which currently equals about 1 mbpd and is likely to grow only very modestly (if at all) over the next 12 years. It also includes government stockpiling. Over the last 10 years, input to strategic stocks has grown by less than 100,000 b/d for the world as a whole. Obviously, any change in government stocking policy would impact this component of crude oil demand.

For global demand growth to average about 1.7 mbpd each year, at least one-third to one-half of that growth will have to come from the Asia–Pacific region. This raises the question of whether the economies of Asia can reach such growth given their current economic problems. Obviously, in the short run they cannot, but in the long-term (after 1999), they are likely to return to a high-growth path by dint of the sheer size of their populations and the ongoing industrialization of their economies. Also, successful deregulation and liberalization of their energy sectors could lower consumer prices and stimulate demand.

The other one-half to two-thirds of global demand is spread throughout the rest of the world, but does include a moderate recovery in refinery utilization in the countries of the FSU, especially after 2005. Capacity utilization in these countries currently averages less than 50 percent; therefore, the potential for higher throughput as the refining sector is refurbished is tremendous, especially in the big refining centers in Russia and Ukraine. The Caspian Sea's role in that growth is discussed below.

Global Crude Oil Supply

In the countries outside of OPEC, crude oil production has grown steadily over the past 15 years, except for the period 1989–93. In the early 1980s, following the Iranian revolution, crude oil prices were high and producing oil was a very profitable business. Non-OPEC output increased by at least 500,000 b/d and often by more than 1 mbpd each year. After the 1986 oil price collapse, the annual rate of growth in non-OPEC output fell under 500,000 b/d. By 1989, the oil sector of the FSU was in a tailspin and for the next five years, due primarily to the implosion of the Russian economy and the dramatic deterioration of the producing industry, non-OPEC production registered at least a 500,000 b/d net decline, annually.

By the mid-1990s, the Russian decline was tapering off and a renaissance had begun in the North Sea. These developments, coupled with new investment in Latin America and Africa, put non-OPEC output back on a 1 mbpd growth track in the latter half of the 1990s. The irony of this most recent oil boom is that it is roughly the size of the early 1980s boom, except that nominal crude oil prices are at least US$10 per barrel lower today than they were 15 years ago, and are even lower given inflation. This comparison highlights the tremendous impact of declining production costs on oil exploration and development.

The supply outlook in this "business as usual" forecast measures non-OPEC supply in over 70 countries worldwide. The basis of the individual country forecasts is both resource potential and access to capital. Steady growth in non-OPEC supplies depends on more than a simple assessment of reserves and production costs. It also depends on the legal and fiscal environment for investment, whether domestic or foreign.

The UK sector of the North Sea is a case in point. In the last decade, the reform of the UK tax regime has dramatically improved the profitability of producing oil in the North Sea. In the early 1980s, government revenues on oil production were drawn from a 12.5 percent royalty on the landed value of the oil produced, a 75 percent Petroleum Revenue Tax (PRT) on the profits from each field, and a 33 percent corporate profits tax. The royalty was abandoned on production from new fields commissioned after 1982 in the Northern Basin and after 1989 in the Southern Basin. In 1993, the PRT was reduced from 75 to 50 percent on existing fields and abolished for all new fields. Even without taking into consideration the many royalty and PRT waivers on existing fields, the marginal tax rate is distinctly lower for newer fields. It is 70.7 percent on a field commissioned before 1982; 66.3 percent on a field commissioned after 1982–89 (depending on location) but before 1993; and 33 percent on a field commissioned after 1993.[2] These tax developments alone have made small, remote and high-cost fields financially viable investments. Now, the UK government is re-thinking these tax breaks, but the recent decline in oil prices seems to have tabled any discussion of their rollback.

The countries of Latin America have very large reserves, but policies regarding foreign investment differ significantly between governments. Argentina has successfully opened its energy sector to foreign investment. Colombia has begun tapping its reserves by allowing foreign companies to enter into joint ventures with its state oil company, Ecopetrol. Brazil is likewise beginning to welcome foreign investment within its new policy of *Flexibilidad*. All three of these countries will be able to expand production with the help of foreign investment, in the form of both capital and expertise. In the case of Mexico, however, continued resistance to foreign investment has and will continue to keep Mexico's production on a slow-growth track, at least between 1998 and 2005. After that, more aggressive development of Mexico's abundant resources could ensue, if policies toward foreign investment change.

In sum, as we look out to 2010 we see two distinct phases of growth in non-OPEC supply. The first is a continuation in the current non-OPEC boom that will peter out around 2000–1. However, due to rising demand, more OPEC output will be needed. As OPEC output increases, the capacity buffer is squeezed and prices will find support. The consequence of moderately higher prices will be renewed interest in the frontier areas of the non-OPEC countries. Higher prices can help finance hidden development costs in countries that have not yet liberalized all of their foreign investment controls. This leads to the second phase of growth in non-OPEC supply, another non-OPEC boom, which yields significant new volumes from the Caspian Basin, Latin America and China, and less significant, but a still important volume of additional barrels from Africa.

The Role of the Caspian

Within the context of non-OPEC supply growth, the Caspian plays an important role. Besides China, the Caspian Basin is seen as the next, and possibly the last, great oil frontier.[3] As a result, even though there is some consensus regarding proven reserves, there is very little regarding ultimate reserves. Since the conclusion of the TengizChevroil deal in Kazakhstan, proven reserve estimates have clustered around 15 to 35 billion barrels. Estimates of total recoverable resources, however, engender much more speculation, ranging from 100 to 200 billion barrels.[4]

Production forecasts for the short term (to 2010) are eerily similar across sources. Many project that production will reach a level of 3.5 mbpd by 2010.[5] This consensus may reflect company and country pronouncements. It may also reflect a simple rule of thumb: at a reserve to production ratio of twenty, 15 to 35 billion barrels equals 2 to 4.8 mbpd, the midpoint of which is 3.4 mpbd. Reaching a reserve to production ratio (r/p) of 20 by 2010 is possible, given the money and technology that has been committed to the region. The TengizChevroil Kazakhstan project and the AIOC offshore Azerbaijan project together will add 1 mbpd to the region's production. As a result, a forecast of 3.5 mbpd, with 1 mbpd from Azerbaijan, 1.8 mbpd from Kazakhstan and the balance from Uzbekistan and Turkmenistan, is optimistic but not out of the question.

The exportable surplus is an entirely different matter. The Caspian currently has about 1.2 mbpd of refining capacity, of which 50 percent is utilized.[6] As these countries grow, in part because of new-found oil wealth, more and more of the spare refining capacity will be refurbished and brought into service to meet local oil demand. At the same time, another nine mbpd of refining capacity in the FSU is only operating at about 45 percent of capacity. Central to this spare capacity is 1.2 mbpd of Ukrainian capacity that is only one-quarter utilized. Obviously, some of the Caspian oil will fill Caspian and other FSU refineries. For the sake of this analysis, however, it is assumed that Caspian crude oil demand will rise to a level commensurate with indigenous refining capacity and that any growth beyond that will be supplied by product imports from the rest of the FSU. Moreover, it is assumed that crude demand in other FSU countries will be met with Russian and not Caspian barrels. This yields a maximum exportable surplus from the Caspian Basin of roughly 2.3 mbpd by 2010, which would represent about 4 percent of total non-OPEC production.[7]

In absolute terms, this is not a large amount, but assuming sufficient port and pipeline capacity to handle it, the impact on price and trade flows by 2010 will be significant. Indeed, of the total *increase* in non-OPEC production from 1998 to 2010, the FSU will account for 40 percent and the Caspian alone will account for 18 percent. By contrast, the North Sea will equal only 4 percent, Latin America 9 percent and Africa 14 percent of the growth in non-OPEC supply. In this sense, it is the *potential* speed more than the potential volume of Caspian oil that is critical to the global oil balance and the "call" on OPEC crude over the next 10 to 15 years.

The Crude Oil Balance

The crude oil balance combines the long-term outlook for non-OPEC crude oil supply, including the Caspian Basin as described above, with the long-term outlook for crude oil demand. In the process, it identifies the volume of crude oil that OPEC will need to put into the market to balance supply and demand. The demand for OPEC oil has been limited in the 1990s as compared to the years following the 1986 price collapse. With the Gulf War and the economic implosion of the Soviet Union, global oil demand grew very slowly in the early 1990s. At the same time, non-OPEC supply (excluding the FSU) underwent a renaissance. As a result, during most of the 1990s, OPEC barrels were not pulled into the market in the same volumes that they were in the late 1980s when lower prices and relative political stability elevated global oil demand. As a result, crude oil prices have been fairly low in the 1990s as OPEC barrels have typically contributed to over-supply.

Indeed, OPEC's power to shape prices has been limited during much of this decade as non-OPEC production increases have largely met demand increases. Contributing to OPEC's impotence has been the rejection of the quota system by Venezuela, which has produced more than 1 mbpd over its quota during the last few years. That over-production, however, finally triggered a Saudi response. In December 1997, Saudi Arabia pushed through the first major quota increase of the 1990s, taking the lion's share of the increase. This move came at a time when non-OPEC production was still growing and Asian oil demand was weakening as a result of the economic crisis. The result was a US$6 drop in oil prices in 1998. The Saudi decision was reminiscent of their 1986 abandonment of the role of swing producer.

During the course of 1998, however, Saudi Arabia and the rest of OPEC have knit together fairly substantial, albeit slow, production cuts to re-balance the market. To date, OPEC has successfully put a floor under prices, but not raised them. However, even with a slowdown in Asian oil demand growth, prices are unlikely to remain at levels that would permanently restrain non-OPEC development. So, in the next couple of years, crude oil prices will remain under the triple pressures of growing non-OPEC output, uncertain OPEC output restraint and slow Asian recovery.

As the next century begins, however, it is increasingly apparent that global demand will again grow faster than non-OPEC supply increases. As a result, for the period 2000–2004, we expect an increase in the call on OPEC crude. This will support prices, which, in turn, will encourage foreign investment in production in all parts of the world. Ironically, this will keep non-OPEC output growing – especially after 2004 – and contain OPEC's share of the growth in oil demand.

The magnitude of the increase in the call on OPEC will depend, in part, on Caspian oil. I present two views of the long-term crude oil balance. In scenario one, Caspian oil development proceeds as described above. The resulting annual increase in the call on OPEC averages about 950,000 b/d for the period 2001 to 2010. In scenario two, the Caspian development does not take place. The resulting annual increase in the call on OPEC averages over 1.15 mbpd. On an annualized average basis this does not seem like a big difference, but by 2010, this means that demand for OPEC crude could reach 41 mbpd. At that level, OPEC is producing at well over 90 percent of its projected productive capacity. With the Caspian, OPEC's output is likely to be only 38 mbpd or less than 90 percent of capacity.[8]

This analysis seems, at first, like an exercise in splitting hairs, but it is precisely these marginal developments that shape global prices. We estimate that *without* Caspian oil supplies, nominal oil prices in 2010 could be as much as US$5 per barrel higher than otherwise. Moreover, this may be a conservative assessment of the price impact. The US Department of Energy (DOE) estimates that every 1 mbpd disruption in supply can increase prices by US$3 to US$5.[9] If the Caspian accounts for 2 mbpd in exports, then, using the DOE's rule of thumb, prices could be as much as US$6 to US$10 per barrel higher. This projection of the price impact, of course, depends on the assumption that no other non-OPEC region in the world could muster the rate of growth that, in theory, the Caspian can.

More important than the absolute price level, however, is the price volatility that results from tighter relative supply. This volatility can exaggerate the impact of temporary supply/demand imbalances and in the process hamper economic growth and slow private investment.

In the search for the Caspian's relevance, there are two quick conclusions. First, among non-OPEC producing regions, the Caspian is unusual in its potential for quick growth. Second, it is precisely this quick growth which makes its impact on marginal non-OPEC supplies profound enough to have a distinct relevance for global oil prices by the end of the next decade.

The price impact described above is based on a "business as usual" supply/demand forecast, which posits a return to fairly strong oil demand growth. It assumes recovery in Asia by 2000 and then sustained global growth for each and every year after that. History has shown, however, that hardly a decade passes without a significant economic slowdown or a political or military event that elevates oil prices, in turn slowing oil demand if not economic growth. Moreover, while the "business as usual" forecast assumes some natural gas replacement of oil, it does not explicitly factor in national legislation that might accelerate that trend in keeping with environmental objectives such as the reduction of CO_2 emissions.

These factors mean that the starting assumption of steady oil demand growth may be misleading. If oil demand does indeed grow more slowly, then the relevance of the Caspian is even greater in contributing to lower prices. Moreover, it would be much harder for OPEC to increase its market share in an environment of lower demand and non-OPEC output increases, 20 percent of which is in the Caspian.

The Impact on Trade Flows and Arab Gulf Producers

One of the most intriguing consequences of Caspian Basin exports will be their impact on trade patterns. As of early 1999, it is not clear by which route(s) the oil will be transported out of the region. The CPC pipeline taking Tengiz oil through Russia to the Black Sea is bound to carry a significant portion of the region's crude. Beyond that route, the next most viable route is unclear. Many believe that swaps with, or transit through, Iran will become an increasingly viable alternative to exporting via Russia, Georgia and Turkey and the Black Sea. Notwithstanding the growing potential for crude oil swaps with Iran and Iran's willingness to invest in Caspian Sea to Arab Gulf infrastructure, American opposition to Central Asian exports coming through the Arab Gulf precludes sizeable volumes following that route, at least in the foreseeable future. As a result, the Baku to Black Sea route is likely to be the second most important route out of the Caspian. In sum, an exportable surplus on the order of 2.3 mbpd (by 2010) and ultimate pipeline and rail capacity of at least that volume indicate that a volume of that magnitude could end up flowing towards the European market.

In assessing Europe's crude oil requirements and alternative suppliers, it seems quite plausible that these barrels will replace some Arab Gulf barrels. This will occur just as Latin American production meets much of North America's growth in import requirements. The result will be that Caspian and Latin American output will meet much of the growth in the Atlantic Basin's crude oil imports. Put another way, Arab Gulf producers will have to compete with favorable pricing to displace these short-haul suppliers.

Another critical factor in assessing the Caspian's relevance in terms of trade flows and market share is the quality of the crude produced. Most of the Caspian oil is light and sweet. Tengiz is very high quality with the American Petroleum Institute's (API) measure of crude oil density or "gravity" in the mid-40s and sulfur under 1 percent, especially when processed to remove hydrogen sulfide and mercaptans.[10]

The Caspian crude appears to be of significantly higher quality than the majority of Arab Gulf exports. In addition, it has a high middle distillate and naphtha yield. Considering that European petroleum product demand will be comprised of cleaner and cleaner automotive fuels, the penetration of Caspian crude into the European market will be encouraged not only by its proximity but its quality.

Expectations for trade between the Arab Gulf and the main consuming markets are summarized in the following tables.

1. *With or Without the Caspian Oil*

   - By 2010, the flow of oil from the Arab Gulf to Asia will rise by at least 6.5 mbpd.
   - By 2010, The flow of oil from the Arab Gulf to North America will fall by about 100,000 b/d.

2. *With the Caspian:*

   - By 2010, the flow of oil from the Arab Gulf to Europe will fall by 1.5 mbpd.

3. *Without the Caspian:*

   - By 2010, the flow of oil from the Arab Gulf to Europe will rise by 0.5 mbpd.

Table 9.2

Main Crude Oil Flows, 1997 (000 b/d)

To

From	Latin America	North America	Asia–Pacific	Europe	Africa	Middle East	FSU	East Europe
Latin America		3500						
North America			250					
Asia–Pacific								
Europe		500						
Africa		1600	1000	2000				
Middle East		1600	9000	3000				
FSU				1500				1000
East Europe								

Table 9.3
Change in Main Crude Oil Flows by 2010 with Caspian (000 b/d)

To

From

	Latin America	North America	Asia–Pacific	Europe	Africa	Middle East	FSU	East Europe
Latin America		2500						
North America			(250)					
Asia–Pacific								
Europe		(500)						
Africa		(600)	500	1500				
Middle East		(100)	6500	(1500)				
FSU			500	1000				500
East Europe								

TABLE 9.4

Change in Main Crude Oil Flows by 2010 without Caspian (000 b/d)

TO

FROM

	Latin America	North America	Asia–Pacific	Europe	Africa	Middle East	FSU	East Europe
Latin America		2500						
North America			(250)					
Asia–Pacific								
Europe		(500)						
Africa		(600)	500	1500				
Middle East		(100)	6500	500				
FSU			500	(1000)				500
East Europe								

Implications for Price

As indicated above, the existence of the Caspian Basin oil and its timely development over the next decade or so will provide a bit more flexibility in the global crude oil balance. By my estimate, this could lead to nominal crude oil prices being lower in 2010 than would have been the case without the Caspian. It is impossible to know how much lower, but the drop is likely to be in the order of US$2 to US$5 rather than US$8 to US$10.

The other important pricing issue is volatility. Since the inception of spot markets, oil prices have been fairly volatile. This is, in part, because oil is an essential energy in almost every country. As such, oil prices are driven by supply and demand developments, but are also influenced by government policies. More recently, as oil is regarded increasingly as a commodity, prices are influenced more and more by the activities of speculators who view oil as just another financial market.

All of these factors can periodically contribute to volatility, but it is probably most extreme in periods when the market is balanced too tight. Then the volatility on the upside can be significant. Indeed, one of the primary benefits of regulated futures markets is that they set daily limits on how much prices can move. This gives the market a chance to stop and assess fast-moving events each day. Even so, price volatility can be an enormous headache for policy makers and investors. To the degree that the Caspian loosens the global balance, it will help temper volatility and contribute to a market that is not easily "cornered."

Implications for Energy Security

During the Cold War, the issue of energy security was clear-cut. Western nations did not want the Soviet Union to gain an advantage over the resources of the Arab Gulf. The primary threat to the flow of oil was Soviet control. This characterization of energy security crystallized with the Soviet invasion of Afghanistan and, shortly thereafter, the adoption of the Carter Doctrine. The Carter Doctrine stated that any attempt by an outside force to control the Arab Gulf would be seen as a threat to the vital interests of the United States and would be repelled by any means, including military force. A secondary concern was political instability within and among the Gulf countries themselves.

A lot has changed since then. The Cold War is over and the perception that the FSU could control oil flows from the Gulf is gone. This has elevated the secondary concern and sharpened the focus on the potential for confrontation within or among the Gulf countries, especially if it affects the flow of oil. As a result, the Carter Doctrine has been reworked into what might be called the Bush Doctrine, or the tacit promise of military assistance in defense against any force that threatens an oil-rich conservative regime in the Gulf. A corollary of the Doctrine is the preference for a UN rather than a US response.

Another development shaping the issue of energy security has been the proliferation of oil-producing countries. Between 1978 and 1996, 22 new non-OPEC countries began producing oil, an increase of more than 40 percent. This is due, in part, to the breakup of the former Soviet Union, Czechoslovakia and Yugoslavia, but it also includes new producing countries in Africa and Asia. Most of these countries have joined the category of producers with production under 50,000 b/d. Over the same period, however, several producers who previously produced below 50,000 b/d have graduated to the 50–100,000 b/d range. Among other producers, there has been a similar "graduation" from the 100–500,000 b/d range to the 500,000 to 1 mbpd size. This latter development is largely the result of impressive growth in output in countries such as Egypt, Oman, Brazil, Argentina, Malaysia, Angola, India, Colombia and Syria. Whether the number of countries that produce oil is growing because of new discoveries in non-producers or because Central Europe and Central Asia have redrawn their borders, it is undeniable that the oil frontier is continuing to grow. Both existing producers and new producers will continue to present opportunities for new exploration and/or production.

With these changes over the last 15 years, the issue of energy security has become less clear-cut. Even though net importing countries are and will remain dependent on oil from the Arab Gulf, the magnitude of the threat seems smaller. In fact, the perception that the threat is less pronounced is shaping behavior and policy. For example, if the United States (with other UN members) is willing to provide a military guarantee for stability in the Gulf, efforts at oil conservation are less urgent. Indeed, it is the environment, not energy security that is now driving the debates over consumption taxes and alternative fuels.

Concern over energy security will never go away, but each new alternative supplier contributes to the perception of a diminishing threat. In this sense, the Caspian does enhance energy security. It is certainly not the "magic bullet" that eliminates the question of energy security, but it represents a volume of oil that is not unimportant as an alternative source.

Since the dissolution of the Soviet Union, the political stability of the Caspian Basin countries has been a concern. The war in Chechnya and the ongoing tensions between Armenia and Azerbaijian over Nagorno-Karabakh have highlighted the potential for political and military conflict in this region. Such conflict is an obvious threat to the flow of oil from the region. It is hard to envision, however, simultaneous political–military crises in the Arab Gulf and Central Asia interrupting oil flowing through the Mediterranean Sea and the Arab Gulf at the same time. Indeed, it is precisely this belief that will encourage opposition to pipeline projects designed to markedly increase the flow of Caspian oil through the Arab Gulf. Ironically, a crisis in one region would probably spur output in the other to take advantage of the windfall afforded by higher prices.

Policies designed to promote energy security are essential. Among those, emergency response measures such as strategic stockpiles and resource-sharing programs should be augmented and perfected. It is precisely the development of regions such as the North Sea, Latin America, Africa and the Caspian, however, that limits the need for anything more.

Relevance for the Global Economy

As for the world economy, the relevance of Caspian oil is that it contributes to a global supply base that will keep oil relatively inexpensive. This can only be good news for the global economy. For those countries whose economies depend on exporting oil, with production costs falling all over the world, producing oil will remain a profitable business even if real oil prices remain flat to falling. For the countries of the Caspian Basin, careful management of new oil wealth will be essential to economic prosperity and political stability. The relevance of the Caspian Basin is summarized in Table 9.5.

TABLE 9.5
Relevance of the Caspian Oil

Price	By adding to supply over the medium and long term, it will temper the upside in prices and marginally lessen price volatility that can create speculative opportunity, but, more often than not, hurts the unwary producer or consumer.
Arab Gulf Producers	Will redraw crude trade patterns, pushing Gulf producers increasingly out of the Atlantic Basin, but into Asia, the region likely to have the greatest growth.
Energy Security	Does not single-handedly remove energy security concerns, but every alternative source of supply lessens those concerns.
Global Economy	Anything that tempers global oil prices, spurs the global economy. In addition, careful management of the export revenues in producing countries could bring about prosperity and political stability.

Conclusions

When discussing the relevance of Caspian oil, one cannot use absolute terms such as "solution" or "panacea." Even so, there are many ways in which the Caspian's oil will re-shape the oil sector as we know it today. Relative to other growing regions, the first significant tranche of Caspian oil will come on line pretty quickly. It will be high-quality oil well positioned geographically to meet the slow growth in demand in Europe. As such, it has the potential to back Arab Gulf (or more correctly growth in Arab Gulf barrels) out of Europe. This will happen while growing crude oil output in Latin America and Canada meets more and more of somewhat faster growth in the US. By 2010, Caspian oil is likely to temper the upside in oil prices, but not to a degree that will slow down non-OPEC development. Any volume that provides a ceiling on the upside in prices, however, will help eliminate OPEC windfalls. It will also serve as an alternative supply, thereby reassuring consuming countries and providing stability for their oil-dependent economies.

Finally, Caspian oil brings to mind the fable of the ant and the elephant. In wandering through the jungle, an elephant comes across an ant stranded on a stick caught in a stream. After the elephant saves him, the ant thanks him and promises to return the favor. The elephant points out that there is nothing a small ant could ever do to help an enormous elephant. They part company and the elephant sets off again through the jungle. During the course of the day, the elephant comes across a lion, then a giraffe and finally a rhinoceros, all of which require his assistance. He helps all three, but each bounds off into the jungle without thanking him. As dusk falls, the elephant himself falls down a steep ravine and becomes wedged in a hole with no chance of escape. Only the tiny ant responds to his call for help. The elephant scoffs when the ant says he can save him. The ant, however, summons hundreds of thousands of his relatives and an army of ants lifts the elephant and carries him to safety.

This story makes two points pertinent to Caspian Basin oil. First, size can be deceiving. In the global oil market, marginal developments in supply or demand can have a disproportionate impact on price and market perception. In percentage terms, expected output from the Caspian Basin is small relative to global supply or global demand, but, as described above, it does not have to be large to have a significant impact on price, trade flows and market dynamics. Second, what may initially seem small and weak can become large and powerful. If the tiny ant can be equated with the 15 to 35 billion barrels of proven reserves in the Caspian Basin, then the 100 to 200 billion barrels of ultimate reserves may one day be the equivalent of an army of ants. Indeed, if proven reserves should inflate to reflect the more optimistic assessments of the region's resources, the Caspian Basin will be one of the most relevant regions in the history of the oil industry.

Contributors

Shirin Akiner

Dr. Shirin Akiner is Lecturer in Central Asian Studies at the School of Oriental and African Studies, University of London. She has extensive first-hand experience in the region and speaks several of the indigenous languages. She has presented papers at a number of international conferences and high-level seminars on Central Asia including in Tokyo, Washington, Ottawa, Geneva, Moscow, Tehran, Tbilisi, Uppsala, Rome, The Hague and London. She is regularly interviewed on Central Asian affairs (e.g. for BBC World Service and Deutsche Welle, as well as French, Dutch, Canadian and Australian networks), and has acted as a consultant on several award-winning radio and television documentaries.

Her authored publications include *Islamic Peoples of the Soviet Union* (London: KPI, 1983, 1987); *Central Asia: New Arc of Conflict?* (London: Royal United Services Institute, 1993); *The Formation of Kazakh Identity* (London: Royal Institute of International Affairs, 1995); and *Minorities in a Time of Change: Prospects for Conflict, Stability and Development in Central Asia* (London: Minority Rights Group, 1997).

Fadhil Chalabi

Dr. Chalabi's professional involvement in oil began in 1968, when he was appointed Director of Oil Affairs at the Ministry of Oil in Baghdad. He was chosen for the office of Permanent Under-Secretary of Oil in 1973, leaving after three years to join OAPEC as Assistant Secretary-General. By 1976, Dr. Chalabi held the office of Deputy Secretary-General of OPEC in Vienna, acting for the Secretary-General himself from July 1983 until June 1988. In 1989, he assumed his present position as Executive Director of the newly established CGES in London, founded and chaired by former Saudi Arabian Oil Minister Sheikh Ahmed Zaki Yamani. Dr. Chalabi received the British Institute of Energy Economics Award for Distinction in 1988.

He is the author of two books: *OPEC and the International Oil Industry: A Changing Structure* (Oxford: Oxford University Press, 1980); and *OPEC at the Crossroads* (Pergamon, 1989). He earned his BA in Law from Baghdad University and a PhD in Economics from the Paris University.

Robert E. Ebel

Robert Ebel is an internationally recognized expert on oil and energy issues at the CSIS in Washington, DC. Formerly Vice-President for International Affairs at Enserch Corporation, Ebel advised the corporation on global issues relevant to day-to-day operations. He has also held posts at the Central Intelligence Agency, the Department of Interior, and the Federal Energy Office. He traveled to the Soviet Union in 1960 as a member of the first US petroleum industry delegation, and in 1971 with the first group of Americans to visit the oil fields in Western Siberia.

Ebel, author of two recent CSIS publications, *Energy Choices in Russia* and *Chernobyl and Its Aftermath*, was project director for Nuclear Power Safety Issues in the former Soviet Union. Ebel earned a BA in Petroleum Geology from Texas Tech and an MA in International Relations from Syracuse University.

Sarah A. Emerson

Sarah Emerson is Managing Director at Energy Security Analysis, Inc. (ESAI), where she currently directs the oil practice. Ms. Emerson oversees all of ESAI's petroleum research, market analysis and price forecasting activities. She has developed many of ESAI's unique analytical tools for analyzing petroleum markets in over 120 countries. In 1992, Ms. Emerson published the first description of the Russian refining sector ever available in the West. In the same year, she organized the first ever translation and publication of former Soviet energy maps.

Ms. Emerson regularly publishes articles in the energy trade press, and co-authored the monograph, *Storage in the International Oil System* (1994) published by the Economist Intelligence Unit, and the book *A Year on the (H)edge: Integrating the Physical and Paper Fundamentals in the Oil Markets* (1995) published by Hobijn French Publishing. Ms. Emerson received her BA from Cornell University and her MA from the Nitze School of Advanced International Studies at Johns Hopkins University.

Paul R. Gregory

Professor Paul R. Gregory is currently Cullen Distinguished Professor of Economics at the University of Houston. He is also a Senior Research Fellow at the Energy Institute of the College of Business Administration of the University of Houston. He served as project coordinator of the Russian Petroleum Legislation Project (1992–95) funded by international energy companies and by the World Bank and the EBRD. He was recently the co-organizer of a major joint energy conference on the former Soviet Union sponsored by Oxford University and the University of Houston in Oxford in March of 1998. His honors include Fulbright, Volkswagen and Humboldt Fellowships. He served as a visiting professor of economics at Moscow State University in 1996.

Professor Gregory has published and researched widely on the Russian and Soviet economics. He is the author of the leading text on the Soviet and Russian economies (now in its sixth edition). He has published more than 10 books, and more than 100 articles on the Russian and Soviet economies and on their energy sectors. Professor Gregory graduated with a PhD in economics from Harvard University.

Amy Myers Jaffe

Energy analyst Amy Myers Jaffe is the Project Coordinator for Energy Research at the James A. Baker Institute for Public Policy at Rice University and a principal author of the Institute's energy studies: *The Political, Economic, Social, Cultural and Religious Trends in the Middle East and the Gulf and Their Impact on Energy Supply, Security and Pricing* and *Unlocking the Assets: Energy and the Future of Central Asia and the Caucasus. A Political, Economic and Cultural Analysis.*

Prior to joining the Baker Institute, Ms. Jaffe served as the Senior Economist and Middle East Analyst for the respected oil journal *Petroleum Intelligence Weekly*. Ms. Jaffe has written for a variety of publications, including the foreign editions of the *Wall Street Journal*, *The New York Times*, *Dow Jones International* and the *Mideast Report* and has appeared on a variety of news shows including CNN, Good Morning America and local broadcasts in New York and Houston. She was the recipient of the 1994 Award for Excellence by the International Association for Energy Economics. Ms. Jaffe is a graduate of Princeton University in Arabic Studies.

Geoffrey Kemp

Geoffrey Kemp is the Director of Regional Strategic Programs at the Nixon Center. He served in the White House during the first Reagan administration and was Special Assistant to the President for National Security Affairs and Senior Director for Near East and South Asian Affairs on the National Security Council Staff. Prior to his current position, he was a Senior Associate at the Carnegie Endowment for International Peace, where he was Director of the Middle East Arms Control Project.

From 1970 to 1980, he was on the faculty of the Fletcher School of Law and Diplomacy, Tufts University. He is the author or co-author of several books on the Middle East, including *The Control of the Middle East Arms Race* (1991), *India and America After the Cold War* (1993), *Forever Enemies?: American Policy and the Islamic Republic of Iran* (1994), *Powder Keg in the Middle East: The Struggle for Gulf Security* (co-editor) (1995), *Strategic Geography and the Changing Middle East* (1997), *Point of No Return: The Deadly Struggle for Middle East Peace* (1997) and *Energy Superbowl: Strategic Politics and the Persian Gulf and Caspian Basin* (1997). He received his PhD in political science at MIT and his MA and BA degrees from Oxford University.

Wilfrid Kohl

A political economist, Dr. Kohl is a Research Professor of International Relations at Johns Hopkins School for Advanced International Studies (SAIS), where he directs the subfield on Energy, Environment, Science and Technology. He is also Director of the International Energy and Environment Program at the Johns Hopkins Foreign Policy Institute. From 1983–85 he directed the University-wide Energy Research Institute. Before coming to SAIS in Washington, Dr. Kohl was director of the SAIS European Center in Bologna, Italy from 1976–80. Professor Kohl has served on the staff of the National Security Council and the Ford Foundation.

He is a member of the International Association of Energy Economics and a regular participant in the Aspen Institute's Energy Policy Forum. Professor Kohl is the author and editor of several books on international relations and energy policy, including *After the Oil Price Collapse: OPEC, the United States, and the World Oil Market* (1991), *Methonal as an alternative Fuel Choice: An Assessment* (1990), *International Institutions for Energy Management* (1983) and *After the Second Oil Crisis: Energy*

Policies in Europe, America and Japan (1982). He earned his PhD in Political Science from Columbia University.

Julia Nanay

Julia Nanay is a Director at the Petroleum Finance Company (Petro Finance), where she provides clients with risk analyses on investments in the oil and gas industry worldwide. Her recent focus is largely on the former Soviet Union, advising on investment opportunities throughout the region. Ms. Nanay has worked in the oil industry in various capacities since 1976. Before joining Petro Finance, she was Vice President of the Charter Oil Company, a Jacksonville, Florida-based refining/marketing company, heading their Washington, DC Office. Previously, she was the Executive Assistant to the Vice President for International Affairs of Northeast Petroleum Industries, Inc., a Boston-based independent marketer.

Over the last year, Ms. Nanay has spoken at many different forums about Russia, Central Asia, the Caucasus, the Caspian region and Iran, including at the Nixon Center; in Ashgabat, Turkmenistan with the Carnegie Endowment; Capitol Hill with the Middle East Policy Council; the Lehrer News Hour; and National Public Radio. She received a BA from the University of California at Los Angeles and an MA in Law and Diplomacy from the Fletcher School of Law and Diplomacy at Tufts University.

Laurent Ruseckas

Laurent Ruseckas, Associate Director for Caspian Affairs at Cambridge Energy Research Associates (CERA), is an authority on politics, economics and energy investment in Central Asia and the Caucasus. Since joining CERA's former Soviet Union Energy practice in 1991, Ruseckas has advised international energy companies on the investment climate and political risk in the Caspian region, traveling frequently to the area. He is the primary author of CERA's semi-annual *Caspian Watch*.

Mr. Ruseckas specializes in several topics related to Caspian energy development, including regional pipeline geopolitics, domestic politics and economic reform in the Caspian states, Russian and Turkish policy toward Caspian energy, and regional security. He is the author or co-author of the following CERA reports: *Oil Investment and the Shifting Sands of Kazakhstani Politics*; *The Unfinished Business of Nagorno-Karabakh: How will it Affect Azerbaijani Oil Development?*; *Kazakhstan's Bold Privatization Program*; *A Tale of Two Seas: The Pipeline Politics of the Caspian and Black Seas*. Ruseckas is a graduate of Harvard College. Since 1995, he has also been a PhD candidate in Political Science at Columbia University, where he is a Jacob K. Javits Fellow.

Notes

1
Caspian Oil Development: An Overview

1 The eight new states were: Armenia, Azerbaijan, Uzbekistan, Georgia, Kazakhstan, Kyrgyzstan, Tajikistan and Turkmenistan.

2 Robert E. Ebel, *Energy Choices in the Near Abroad: The Haves and the Have-nots Face the Future* (Washington, DC: CSIS, 1997); Rosemarie Forsythe, *The Politics of Oil in the Caucasus and Central Asia: Prospects for Oil Exploitation and Export in the Caspian Basin* (London: International Institute for Strategic Studies, 1996); Sheila Heslin, "Key Constraints to Caspian Pipeline Development: Status, Significance, and Outlook." James A. Baker Institute for Public Policy (April 1998).

3 Richard L. Morningstar, "Testimony before the Senate Subcommittee on International Economic Policy, Exports, and Trade Promotion." March 3, 1999; US Department of State, "Caspian Region Energy Development Report as Required by HR 3610." Congressional Record, May 8, 1997.

4 Rajan Menon, "Treacherous Terrain: The Political and Security Dimensions of Energy Development in the Caspian Sea Zone." *NBR Analysis* vol. 9, no. 1 (February 1998).

5 Amy Myers Jaffe and Robert Manning, "The Myth of the Caspian 'Great Game': The Real Geopolitics of Energy," *Survival* vol. 40, no. 4 (Winter 1998): 112–131.

6 Ebel, op. cit.; Geoffrey Kemp, *Energy Superbowl: Strategic Politics in the Persian Gulf and the Caspian Basin* (Washington, DC: Nixon Center for Peace and Freedom, 1997); Jaffe and Manning, op. cit.

7 US Department of State, op. cit.

8 John Roberts, *Caspian Pipelines* (London: Royal Institute of International Affairs, 1996).

9 "Early oil" refers to the oil produced from an existing platform in the AIOC contract area (the Chirag-I platform), which will ultimately be able to produce about 130,000 b/d. Production beyond that level ("main oil") will require the construction of new platforms at different locations within the contract area.

10 Morningstar, op. cit.

11 Kemp, op. cit.; Frederick Starr, "Power Failure: American Policy in the Caspian." *National Interest* no. 47 (Spring 1997): 20-31.

12 South Pars is a very large offshore gas field in the Arab Gulf which is being developed by Iran in cooperation with a consortium consisting of Total, Gazprom and Petronas. Under pressure from the EU, the United States government granted a waiver to the Iran-Libya Sanctions Act to prevent the project from triggering US sanctions against the participating countries.

2
DEVELOPING CASPIAN ENERGY RESERVES: THE LEGAL ENVIRONMENT

The author wishes to thank Gurcan Gulen, CBA Energy Institute, University of Houston; Gary Conine, Law Center, University of Houston; Christian v. Hirschhausen, DIW Berlin; Joseph Hakkinen and Fred Grasso for their useful comments on an earlier draft of this chapter and also Zalina Tsarakova for her research assistance.

1 Graham Fuller, "Geopolitical Dynamics of the Caspian Region," *Caspian Crossroads Magazine* vol. 3, no. 2, Fall 1997, 1 (electronic version, EV).
2 These estimates are from Bhamy Shenoy and Gurcan Gulen, 4–5. They are drawn from estimates of the Energy Information Administration, *Oil and Gas Journal, BP Statistical Review of World Energy*, and *Interfax Energy Projects*.
3 These figures are from Shenoy and Gulen, "Caspian Pipeline Game," Energy Institute, College of Business Administration, University of Houston, 1998.
4 Margaret Coker, "Agreement Could Unlock Caspian," CBS MarketWatch, July 7, 1998.
5 Daniel Yergin and Thane Gustafson, "Evolution of an Oil Rush," *New York Times*, August 6, 1997.
6 According to Oxman: "It is necessary to have some idea of what we mean by 'sea' in this context. If we mean semi-enclosed marine areas subject in their entirety to the sovereignty of coastal states, then there are two possibilities generally recognized by the Law of the Sea: juridical bays surrounded by a single state whose entrance does not exceed 24 miles or, in relatively rare instances, historic bays or waters over which coastal states have acquired title by prescription over a long period of time . . . None of these examples fits the Caspian Sea, at least not perfectly. It is not a juridical bay because it is surrounded by more than one state. It is automatically subject to the control of the riparian states . . . Unlike semi-enclosed marine seas of comparable size, It is subject in its entirety to the sovereignty of the riparian states." Bernard Oxman, "Caspian Sea or Lake: What Difference Does It Make?'" *Caspian Crossroads Magazine* vol. 1, no. 4 (Winter 1996).
7 The former Soviet states are Kazakhstan, Azerbaijan, Turkmenistan, Uzbekistan, Tajikistan and Kyrgystan. Other states are Afghanistan, Iran, Pakistan and Turkey. These ten countries form the ECO, which is an international political organization of Islamic countries. For a description, see "Ten Caspian States Meet and Talk," *Alexander's Oil and Gas Connections*, May 11, 1998. With the growing importance of Georgia as a transit point, Georgia should also be included as a player.
8 Fuller, op. cit.

9 According to Fuller: "As we try to think of where Russia is headed in the future, there are major unknowns, both domestic and foreign, that will determine how Russia will act in this region. Russia's relations with the West are the key. Will they be perceived to be adversarial or not? If they are, then perhaps we will see a nasty effort by Russia to create neo-imperialistic control over its former republics. I am hopeful because in the last couple of years, Russia has been showing definite signs of progress towards pragmatism and common sense. Of course, there is a long way to go, but I am nonetheless encouraged by the direction of its progress." Fuller, ibid, 2 (EV).

10 Since 1995 Russia has refused to transport Turkmen gas to Western Europe, diverting it instead to the former Soviet republics. At the start of 1997, Turkmenistan ceased delivery of gas to these states due to payment problems. Russia and Turkmenistan have yet to reach agreement on transit tariffs (Deutsches Institut Fuer Wirtschaftsforschung, DIW).

11 "Iran Attacks Caspian Oil Deal," BBC News Online Asia–Pacific, July 8, 1998.

12 According to Oxman: "Unlike the sea, rivers and lakes are subject to national sovereignty . . . and their use for navigation or any other purposes is subject to authorization and control by one or more riparian states . . . Rivers and lakes normally come under state sovereignty by virtue of the prolongation of sovereignty from land; the right to regulate activities in rivers and lakes that border more than one state is usually allocated geographically between the riparian states by agreement and is also subject to special regimes designed to accommodate the interests of all the riparian states in navigation, non-navigational uses and environmental protection." Oxman, op. cit., 2 (EV).

13 Azeri President Aliyev has been the most vocal proponent of demilitarization of the Caspian as a means to resolve the question of navigation rights.

14 A major portion of the Caspian Sea sturgeon stock spends the summer months in the northern part of the Sea (and the Volga and Ural rivers) and the winter months in the central and southern parts of the Sea. This poses conservation and management problems. Geographic partition cannot ensure sound management or equitable allocation of migrating stocks. The choice is between cooperative arrangements among the riparian states based on flag state jurisdiction and cooperative arrangements based on riparian state jurisdiction, or some combination of the two.

15 Oxman, op. cit., 13 (EV).

16 As demonstrated by the case of Lake Constance, it is possible to agree on the regulation of particular uses and achieve pragmatic accommodations, even between neutrals and belligerents in time of war, without resolving the question of whether there is to be a condominium or partition and therefore without any formal and complete partition of sovereignty. An exception appears to be the Gulf of Fonesca. A chamber of the International Court of Justice found the Gulf of Fonesca to be an historic bay that would have had the status of internal waters as a juridical bay had it been bordered by only one state. The Chamber found that each of the three riparian states was entitled to a three-mile belt off its coast, but beyond that found the waters of the Gulf to be held in common by the three states. In explaining its conclusion, the Chamber stated: "[T]hose waters were waters of a single-State bay during the greater part of their known history. They were, during the colonial period and even during the period of the Federal

Republic of Central America were not divided between the different administrative units which at that date became the three coastal States of El Salvador, Honduras and Nicaragua. There was no attempt to divide and delimit those waters" (Oxman, op. cit., 5, EV).

17 In semi-enclosed seas, beyond the maximum 12-mile limit of the territorial sea, the coastal state exercises sovereign rights over the exploration and exploitation of mineral resources within the 200-mile exclusive economic zone and over any continental margin that extends beyond that limit. For example, the Law of the Sea Convention establishes detailed rules regulating the location of offshore installations and safety zones around them to prevent accidents. If the regime in a lake is such that no foreign navigation or other activity occurs or is permitted in areas where the riparian state has established drilling rigs or other installations, then it is up to that riparian state to decide the extent, if any, to which it wishes to adopt policies to protect its own ships and routes (subject to its environmental obligations). But if the regime in a lake is such that foreign navigation, fishing or other activity exists as a matter of right or is otherwise permitted in the relevant areas, then some accommodation of uses is required and the rules of the Law of the Sea are likely to provide a useful guide. (Oxman, ibid, 7–8, EV).

18 A comprehensive boundary treaty of 1954 delimited the boundary on both sides of the Caspian but was silent as to any boundary within the sea itself. Earlier treaties were likewise silent as to a sea boundary.

19 When the area containing the Azeri field was tendered for bid in January 1991, the tender was made jointly by the USSR Ministry of the Oil and Gas Industry and by the Council of Ministers of Azerbaijan. A similar procedure was applied for international tenders in Kazakhstan.

20 Friedeman Mueller, "Die Region des Kaspischen Meeres Energiereichtum und Geopolitik," *Osteuropa Wirtshaft*, vol. 41, no. 3 (1996)

21 Jamshid Momtaz, "The Legal Regime of the Caspian Sea," *Amu Darya* vol. 1, no. 2 (Summer/Fall 1997).

22 "Irate Iranians," *Russian Petroleum Investor*, December 1997/January 1998, 46.

23 Richard Pomfret, "The Economic Cooperation Organization: Current Status and Future Prospects," *Europe-Asia Studies* vol. 49, no. 4 (June 1997).

24 A.H. Hosseini, "The Prospects for the Caspian Sea Region in the 21st Century", Institute for Political and International Studies, (Summer/Fall) 1997.

25 Ibid.

26 Examples of similar treaties include those between the United States and Canada, the Clean Caribbean Cooperative involving the United States, Trinidad and Tobago and other Caribbean countries; the Kuwait Regional Convention for Cooperation on the Protection of the Marine Environment from Pollution and a host of others. On this, see Brice Clagett, "Rules of International Law," *Caspian Crossroads Magazine* vol. 1, no. 9, December 1996, 4–5, EV.

27 Clagett notes that: "While states are free to agree on any boundary they wish, the overwhelming majority of treaties that delimit boundaries in lakes or inland seas have adopted the median/equidistance line between the opposite shores, occasionally modified to take account of some historic or other special circumstance. Notably, the Great Lakes of North America – whose size makes them the bodies of water most closely analogous to the Caspian – were delimited

by equidistance as long ago as 1783. A few of the many other examples include Lakes Geneva, Tanganyika, Lugano and Maggiore. The law is also clear that internal administrative boundaries become 'transformed into international frontiers' when a state breaks up into multiple successor states. A tentative boundary could be modified to avoid dividing an established resource deposit, especially if that deposit has a historic association with one of the parties." Clagett, ibid.

28 Andrei Shoumikhin, "New Developments Related to Caspian Oil," Proceedings of the International Workshop "Caspian Oil: Transcaucasus, Central Asia and International Security," Moscow Public Science Foundation, vol. 1, no. 9, December 1996, 1.

29 DIW.

30 "Ten Caspian States Meet and Talk," *Alexander's Gas & Oil Connections*, May 11, 1998.

31 "Turkmenistan and Azerbaijan Clash Over Caspian Sea," *Alexander's Gas & Oil Connections*, January 30, 1997.

32 "Back from the Dead," *Russian Petroleum Investor*, February 1997, 28.

33 Consider this statement of the Embassy of Turkmenistan: a statement by the Ministry of Foreign Affairs, July 7, 1998 Washington, DC. "A Russian–Azerbaijan Agreement on the joint exploration of the Caspian, with the participation of the State Oil Company of Azerbaijan Republic, Lukoil and Rossneft of Russia, was signed on July 4 in Moscow. While such efforts of cooperation are welcome by Turkmenistan, this particular agreement includes joint exploration of Turkmenistan's 'Serdar' deposit (formerly known as Intermediate) and appears in the agreement as Azerbaijan's 'Kyapaz' field. Turkmenistan has voiced its support for establishing an international legal order in the Caspian Basin, based on the new political realities, the interests in developing the deposits and the interests of all the littoral states. To this end, Turkmenistan has supported calls, including those from Moscow, for restraint in proceeding with physical exploration and development of the Caspian Sea prior to an internationally-recognized legal status. Turkmenistan objects to the flagrant disregard of its legal rights to the 'Kaverochkin' and '26 Baku Commissars,' which have been claimed and renamed 'Chirag' and 'Azeri' respectively by Azerbaijan. The Russian Federation and other littoral states are aware of Turkmenistan's opposition to these claims. Turkmenistan has repeatedly asked the Azeris to refrain from proceeding with development of the fields and to enter into constructive dialog regarding the two sides' claims . . . The Turkmenistan government has authorized an official protest to the above mentioned agreement."

34 "Azerbaijan, Russia Discuss Caspian Sea Status," *Itar-Tass*, March 27, 1998.

35 "Much Ado About Nothing," *Russian Petroleum Investor*, April 1998, 32.

36 "Russian oilmen are used to thinking big, within the limits of the former USSR, not Russia alone," said an oil analyst at the United Financial Group: "What they really needed were some rules of the game in the Caspian. Now they've got basic rules."

37 The Iranian protest relates to the launching ceremony of Azerbaijan's production from the Chyrag field, November 10, 1997.

38 "Irate Iranians," *Russian Petroleum Investor*, December 1997/January 1998, 46.

39 "Iran Increasingly Isolated on Caspian Argument," *Alexander's Gas & Oil Connections*, May 4, 1998.

40 "Cross Caspian Pipeline Would Violate Rights," *Alexander's Gas & Oil Connections*, April 23, 1998.
41 Apparently Iran itself is not united on this issue. Iran's ministry of oil has criticized the foreign ministry for its actions which have caused Iran losses in the region by acting too late ("Iran Increasingly Isolated," op. cit.).
42 Oxman, op. cit., 7–8 (EV).
43 Ibid.
44 Shoumikhin, op. cit.
45 "Caspian Geological Peculiarities Need to be Taken Into Account," *Alexander's Gas & Oil Connections*, April 23, 1998.
46 "Azerbaijan, Russia Discuss Caspian Sea Status," *Itar-Tass*, March 27, 1998.
47 ConocoEurasia, "The Current Political and Economic Challenges in Azerbaijan, Georgia, Turkmenistan and Kazakhstan" July 1998.
48 World Bank, Strategic Export Expansion Study, *Interim Report*, July 22, 1996.
49 European Commission Directorate-General XVII (Energy), The Energy Charter Treaty, September 15, 1997.
50 The ECT's basic aim is to create a legal framework to promote long-term cooperation in the energy sector, including exploration, production, transit and trade, the protection of investments and the transfer of profits. The ECT provides for stringent procedures to settle disputes between states as well as between states and companies. Trade in energy materials and products between signatories shall be governed by the provisions of the General Agreement on Tariffs and Trade (GATT) and its related instruments, even if one of them is not a member of the World Trade Organization WTO. The enforceability of the ECT is ensured through dispute settlement mechanisms, providing for final and binding solutions for disputes between states and those between states and investors. Not all provisions will fully apply to all signatories immediately after ratification and coming into force of the ECT. Countries with economies in transition have limited transitional arrangements. The Energy Charter Conference, established by the Treaty, will monitor compliance with the ECT's provisions and act as a forum for East–West dialogue on energy issues.
51 Currently, relatively small amounts of Caspian oil are being exported through Georgia and through swaps with Iran.
52 "Progress for the CPC Does Not Necessarily Mean Progress for Oil Producers in Kazakhstan," *Russian Petroleum Investor*, June/July 1998.
53 In May 1997, the ECO states signed a declaration "On the development of Transportation and Telecommunications Infrastructure and the Network of Transnational Pipelines" which expresses ECO members' moral and political support for two pipeline projects (via Afghanistan and through Iran to Turkey). The latter was signed by President Niyazov of Turkmenistan and the prime ministers of Iran and Turkey ("Planned Pipelines Would Open Turkmenistan," *Russian Petroleum Inventor*, June/July 1997). On May 12, 1998, President Nazarbayev announced his country has interests in several oil transportation projects within ECO and that cooperation within ECO is very important. He named the CPC pipeline from Tengiz to Novorissisk as one of the pipeline projects that requires such cooperation ("Kazakhstan Betting on Several Horses," *Alexander's Gas & Oil Connections*, May 12, 1998).
54 Jennifer DeLay, quoted in *Russia Today*, June 8, 1998.

55 In December of 1997, for example, government officials from Iran, Kazakhstan and Turkmenistan met in Tehran to discuss the proposed pipeline from Kazakhstan and Turkmenistan through Iran. This same project had been the object of discussion in an ECO meeting in Ashgabat in November. On this, see "Unclogging the Pipe," op. cit.

56 Shenoy and Gulen, op. cit., 5.

57 "ECO Foreign Ministers Meet," *Alexander's Gas & Oil Connections*, May 11, 1998.

58 This discussion focuses only on the Caspian states that were part of the Soviet Union. Iran's national legal regime is not discussed.

59 ECT signatories are to encourage and create stable, equitable, favorable and transparent conditions for foreign investment through the better of most favored nation or national treatment. For the pre-investment phase, however, the national treatment principle will be implemented in two stages. In the first stage, following the 1994 ECT, it will be granted on a "best efforts" basis. In the second stage, it will be granted subject to conditions to be defined in a supplementary treaty.

60 "Deeper Pockets," *Russian Petroleum Investor*, December 1997/January 1998, 41.

61 Thane Gustafson, *Crisis Among Plenty: The Politics of Soviet Energy Under Brezhnev and Gorbachev* (Princeton, NJ: Princeton University Press, 1989).

62 Quoted in Paul Gregory, *Reforming the Soviet Economic Bureaucracy* (Cambridge: Cambridge University Press, 1983).

63 Thomas W. Waelde and Christian von Hirschhausen, "Regulatory Reform in the Energy Industry of Post-Soviet Countries," *DIW Discussion Paper* no. 155, November 1997.

64 The Constitution of the Russian Federation was passed in a nationwide referendum on December 12, 1993. The Russian Constitution is the highest law of the land, superseding all other laws or regulations in the case of conflicts. The Russian Constitution guarantees the right of citizens to own private property and to use that property to engage freely in economic activity. These same rights and others relating to the freedom to engage freely in economic activity are also supported in the Civil Code (*Grazhdansky Kodeks*) passed on October 21, 1994, which defines juridical persons and sets their rights equal to those of private persons, and also provides procedures for the registration of companies and their liquidation, or bankruptcy. It also establishes the rights and legal forms of partnerships, corporations and cooperatives. The Civil Code defines the legal foundations for corporations (closed or open), states how they are to be formed, charter capital, the emission of shares and their reorganization or liquidation. The Civil Code establishes the ruble as the sole legal currency and the right of citizens to hold foreign currency. The general framework of legislation of the Russian Federation formally establishes a legal environment for a market economy. The legislative framework, however, is only part of the process of creation of legal institutions. Each separate piece of legislation usually calls for a whole series of regulations, or as the Russians say "normative acts" which provide the specifics behind the more general legislation. The Russian Law on Underground Resources was enacted by parliament in February of 1992 and signed by President Yeltsin in May of 1992 as the umbrella law covering all underground resources. The Law on Underground Resources, in keeping with the later-enacted Russian Constitution, divided ownership of underground resources between federal, state and local jurisdictions

and spelled out provisions for licensing of the development of mineral resources. Although Russian authorities intended to have a petroleum law to govern issues such as ownership, licensing and transportation, such a law was never enacted. In place of a general law on oil and gas, the Russian government and parliament have focused on licensing provisions and on working out the operational specifics of production sharing arrangements. The Production Sharing Agreement (PSA), signed on December 31, 1995 but not approved in its final form, defines the relationship between the state and oil companies including foreign oil companies. PSAs have been used by international oil companies to operate in numerous foreign countries. They spell out the manner in which production from a particular venture is shared between the companies and the state and how these will be distributed over time. For a discussion, see Gregory, op. cit., 13–15.

65 "Pro-PSA Deputies Plan to Enact PSA Laws This Summer," *Russian Petroleum Investor*, June/July 1998, 18–19.

66 "For Three Joint Ventures, Tax Liability Expands Beyond Expectations," *Russian Petroleum Investor*, June/July 1998, 22–5.

67 The Foreign Investment Law of 1994 provides 12 governmental guarantees for foreign investors and exemptions from certain customs duties. The Oil Law sets forth preferred methods for awarding contracts for petroleum exploration and production rights through a competitive tender process. The Oil Law provides for private ownership of oil after it has been lifted. On this, see Daniel Slone and John Lain, "Open For Business: The Legislative Framework of the Republic of Kazakhstan," *Caspian Crossroads Magazine* vol. 1, no. 3 (Summer/Fall 1995): 1–4, EV.

68 "Caspian Investors Contend with Home-Grown Rival," *Russian Petroleum Investor*, June/July 1997, 32–4.

69 "KazakOil's Ambitious Growth Goals Offer New Opportunities to Investors," *Russian Petroleum Investor*, June/July 1998, 29–31.

70 Shirin Entezari and Aleen Rothschild-Seidel, "Azerbaijan: Laws Concerning Privatization and Foreign Investment," *Caspian Crossroads Magazine* vol. 1, no. 3 (Summer/Fall 1995), 1–4, EV.

71 Denton Hall, Doing Business in Azerbaijan (London: Denton Hall International Group, 1998)

72 When confronted with rising construction costs, foreign consortium members have been confronted with demands by Azeri officials to absorb the extra costs ("Baku's New Financial Demands"). For example, in May 1998, the Azeri President (Aliyev) demanded that the AIOC foreign partners allow it to decide on the choice of pipeline routes and to pay for additional pipeline construction costs that can emerge.

73 "Better Than Nothing," *Russian Petroleum Investor*, April 1997, 47.

74 "Economic, Not Political, Reforms Are More Likely to Ensure Contract Stability in Turkmenistan," *Russian Petroleum Investor*, June/July 1998, 33–4.

75 C. Hirschhausen and H. Engerer, "Gas Sector Restructuring in the CIS: A Political Economy Approach," DIW Berlin, February 1998.

76 "Foreign Investors May Finally get Access to Turkmenistan's Offshore Deposits," *Russian Petroleum Investor*, June/July 1997, 39–40.

77 Quoted in "The Perfect Host," *Russian Petroleum Investor*, April 1998, 41.

78 The EBRD gives the non-Russian Caspian states some of its worst "corruption index" rankings.
79 *Russia, Kazakhstan Ink Caspian Sea Deal,* press release, July 7, 1998.
80 "Russia. Kazakhstan Sign Oil Pact, Ending Protracted Row," BT, July 7, 1998.
81 For a recent example relating to Kazakhstan, see "CCL Wins a Round in Court in its Battle with Astan, but the Fight Continues," *Caspian Investor,* September 1998, 29–33.
82 An example of potential problems in this area is the dispute over the Russian-appointed manager of the CPC pipeline, Vladimir Stanev, who had to be removed by Western consortium partners. On this, see "Pipeline Dreams: The Struggle for Caspian Oil," *Washington Post,* October 6, 1998.
83 ConocoEurasia, The Current Political and Economic Challenges in Azerbaijan, Georgia, Turkmenistan and Kazakhstan, July 1998.
84 Ibid.
85 Russia and Azerbaijan dispute an offshore field off the Russian Daghestan coast called D222-Yalama.
86 A well-known case of investment losses due to territorial disputes (involving disputes between Iran and Umm Al Qaiwain and Sharjah) is discussed in *Occidental of Umm Al Qaiwain, Inc* v. *A Certain Cargo of Petroleum,* 577 F.2d 1196 (5th Circuit, 1978).
87 These figures are from the DIW database, DIW Berlin.

3
IRAN AND CASPIAN ENERGY: PROSPECTS FOR COOPERATION AND CONFLICT

1 Frank Viviano, "Caspian Basin Oil Boom Transforms Entire Area," *Washington Times,* August 15, 1998, A6.
2 Martha Brill Olcott, "The Caspian's False Promise," *Foreign Policy* no.111 (Summer 1998): 96.
3 Marshall DeLuca, "Caspian Mobile Rig Clubs pick up Slack in Availability," *Offshore,* July 1998, 68, 70.
4 Over the years, the major oil- and gas-exporting countries have included: Algeria, Angola, Bahrain, Brunei, Canada, Colombia, Egypt, Iran, Indonesia, Iraq, Kuwait, Libya, Mexico, Nigeria, Norway, Qatar, Oman, Romania, Russia, Saudi Arabia, United Arab Emirates, the United Kingdom, the United States and Venezuela.
5 John C. Colombos, *International Law of the Sea* (London: Longman, 1961), 164.
6 "Iranian Ambitions in Caspian Region Eyed," Moscow Aziya I Afrika Segodnya, no.12, December 1994, 15-18, in FBIS-SOV, 10 March 1995, 1.
7 Clive Schofield and Martin Pratt, "Claims to the Caspian Sea," *Jane's Intelligence Review,* February 1996, 77–78.
8 Terence Adams and Gregory Rich, "Greater Power Politics and the Azerbaijan Pipeline: An Update," *Policywatch* (Washington, DC: Washington Institute for Near East Policy, February 24, 1997).
9 Ibid.

10 Elaine Sciolino, "It's a Sea! It's a Lake! No. It's a Pool of Oil," *New York Times*, June 21, 1998, 16

11 Kelly Couturier, "Turkey Aims to Satisfy Its Fuel Needs," *Washington Post*, October 20, 1997, A17–18.

12 "Khatami Calls for Expansion of Iran-Georgia Cooperation," *Xinhua News Agency*, June 14, 1998.

13 "Cooperation Agreement Signed at end of Georgian Foreign Minister's Visit," *BBC Summary of World Broadcasts/Vision of the Islamic Republic of Iran Network 1*, June 23, 1998.

14 Ahmed Rashid, "The Turkmenistan–Afghanistan–Pakistan Pipeline" (Washington, DC: Petroleum Finance Company, September 1997), 10.

15 Anna Kurbanova and Lyudmila Glazovskaya, "Turkmen–Iranian Relations Important Factor in World Policy," *TASS News Agency*, July 8, 1998.

16 "Turkmenistan, Iran Sign Joint Communiqué, Agreements," *BBC Summary of World Broadcasts/IRNA*, July 11, 1998.

17 Ibid.

18 "Survey of Central Asia," *The Economist*, February 7, 1996, 12.

19 "Kazakhstan to Develop Ties with Iran Despite US Objections," *BBC Summary of World Broadcasts/TASS News Agency*, November 15, 1997.

20 "Irani, Kazakh Firms Discuss Oil Cooperation," *Xinhua News Agency*, March 14, 1998.

21 See four-part series in the *Financial Times*, July 7, 1998, including: Robert Corzine and Robin Allen, "Deals Shaped with an Eye to Constitution," 4; Robert Corzine, "Tehran Reopens Doors to Vast Oil and Gas Resources," 4; "Top Companies Lured by Huge Opportunity," David Gardner, "US Action Underlines Investment Needs," 4.

22 Eliyahu Kanovsky, *Iran's Economic Morass: Mismanagement and Decline Under the Islamic Republic* (Washington, DC: Washington Institute for Near East Policy, 1997), Appendix 4.

23 Elaine Sciolino, "Iran's Difficulties Lead Some in US To Doubt Threat," *New York Times*, July 5, 1994, A1.

24 "Iran: Economic Structure;" "Iran: Quarterly Indicators of Economic Activity," EIU Country Reports, November 17, 1996.

25 Laurent Ruseckas and James Placke, "US Policy and the Caspian Pipeline Derby: Dynamism Amid Dilemmas," report by CERA, July 1998; and, Ronald Soligo and Amy Myers Jaffe, "Unlocking the Assets: Energy and the Future of Central Asia and the Caucasus – A Political, Economic, and Cultural Analysis," report on the Economics of Pipeline Routes: The Conundrum of Oil Exports from the Caspian Basin (Baker Institute for Public Policy: Rice University), April 1998.

26 Federico Peña, Secretary of Energy, statement before the Committee on International Relations, US House of Representatives, April 30, 1998.

27 Ibid.

28 The Center for International Political Economy and the Baker Institute for Public Policy, "Unlocking the Assets: Energy and the Future of Central Asia and the Caucasus" (Houston: Rice University, April 1998).

29 "Countries Delay Caspian Sea Agreement, IISS Questions Oil Potential of Region," *The Energy Report* vol. 26, no. 18, May 4, 1998, 95–113.

30 See Olcott, op. cit.

31 David Filipov, "US Warns against Pipeline Going through Iran, but Few Listen; Race to Export Turkmenistan's Gas Looks to Logical Route," *The Boston Globe,* March 1, 1998, A12.
32 Ibid.

4
POLITICAL TRENDS IN THE NEW CASPIAN STATES: AZERBAIJAN, KAZAKHSTAN AND TURKMENISTAN

1 Vote by the Azerbaijan Supreme Soviet. However, this was not formally ratified until October 17 and this date is now celebrated as Independence Day.
2 The Belovezhsky Agreement concluded on December 8, 1991 between the leaders of the three Slav republics marked the end of the Soviet Union.
3 In the all-Union referendum on the future of the Soviet Union, held on March 17, 1991, Turkmenistan returned a vote of over 94 percent in favor of remaining within the Union.
4 Most of the commentary on the war in Karabakh, while purporting to be objective, tends to favor either the Armenians or the Azerbaijanis. For instance, compare the report by Human Rights Watch/Helsinki, *Azerbaijan: Seven Years of Conflict in Nagorno-Karabakh* (New York, NY: Human Rights Watch, 1994), with Levon Chorbajian et al. *The Caucasian Knot: The History and Geo-Politics of Nagorno-Karabagh* (London: Zed Books, 1994).
5 The problems of Kazakhstan are well documented in Yelena Kalyuzhnova, *The Kazakhstani Economy: Independence and Transition* (London: Macmillan, 1998). For an overview of economic trends in all three countries see EBRD, *Transition Report 1998* (London: EBRD, 1998).
6 The hydrocarbon resources of the Caspian Sea have already spawned a very considerable literature. Some of the most useful works are Rosemarie Forsythe, *The Politics of Oil in The Caucasus and Central Asia: Prospects for Oil Exploitation and Export in the Caspian Basin,* Adelphi Paper 300, IISS (Oxford: Oxford University Press, 1996); John Roberts, *Caspian Pipelines* (London: Royal Institute of International Affairs, 1996); Ottar Skagen, *Caspian Gas* (London: Royal Institute of International Affairs, 1997); Institute of Developing Economics, (IDE) Spot Survey, *The Caspian Basin Oil and Its Impact on Eurasian Power Games,* edited by Manabu Shimizu (Tokyo: Institute of Developing Economies, June 1998). See also other chapters in the present book on related issues.
7 Aliyev ran against two virtually unknown candidates. He gained 98.8 percent of the vote. Elizabeth Fuller, "Azerbaijan at the Crossroads," in Roy Allison (ed.) *Challenges for the Former Soviet South* (London: Royal Institute of International Affairs, 1996), 129, pointed out that "Some observers attributed Aliyev's political comeback to interference by Moscow . . ." A concise chronicle of Aliyev's return to power is given by Audrey Altstadt, "Azerbaijan's Struggle towards Democracy," in Karen Dawisha and Bruce Parrott (eds) *Conflict, Cleavage and Change in Central Asia and the Caucasus* (Cambridge: Cambridge University Press, 1997), 124–29.

8 He gained 76.11 percent of the vote. However, members of the opposition disputed the official election figures and claimed that one of the other candidates, Etibar Mamdedov, Chairman of the Azerbaijan National Independence Party, gained over 35 percent of the vote (Richard Giragosian, *TransCaucasus: A Chronology*, vol. 8, January 1999 [Washington DC: Armenian National Committee of America, EV]).

9 Ermukhamet Ertysbayev, "Nekotoryye aspekty prezidentskoy izbiratel'noy kampanii v Kazakhstane," *Tsentral'naya Aziya i Kavkaz* vol. 1 no. 3 (1999): 44–55, states that "almost 99 percent of Kazakhstanis" voted for Nursultan Nazabayev in December 1991. Amendments to the Law on Elections were introduced into the Constitution in October 1998; these included the extension of the presidential term of office and the lifting of the age limit on presidential candidates. In 1999 three alternative contenders stood against Nazarbayev, but, as Ertysbayev indicates, they were not regarded as serious (or, in his words, "legitimate") candidates. All the genuine rivals to Nazarbayev were excluded on various procedural grounds. Nazarbayev received 79.78 percent of the vote.

10 Turkmenistan was the first member of the CIS to take the decision to prolong the term of office of the incumbent president by referendum. Uzbekistan followed suit the following year.

11 See notes 9 and 10 above. In Azerbaijan, too, amendments were introduced into the constitutions in 1998 that facilitated the re-election of the incumbent president. (See El'bar Ismailov, "Vlast' i oppozitsiya nakanune i v period prezidentskoy izbiratel'noy kampanii v Azerbaydzhane," *Tsentral'naya Aziya i Kavkaz* vol. 1, no. 2 (1999): 33–43.)

12 The most ostentatious monument is the 75-meter arch, supporting a gold plated 12-meter statue of President Niyazov, with arms raised toward the sun, that was unveiled in January 1999 to commemorate the third anniversary of Turkmenistan's Declaration of Neutrality. Situated in the center of Ashgabat, the capital, the arch is intended to symbolize a traditional Turkmen hearth; it has a café on its upper tier. The statue of the president rotates, so that it is always facing the sun.

13 "Insulting the honor and dignity of the President" is a charge that is not infrequently invoked in Azerbaijan; see, for example, cases cited in Ulvi Hakimov, *Azerbaijan Bulletin-57*, December 23, 1998, Azerbaijan National Democracy Foundation, Baku. In Kazakhstan, too, law protects the honor of the president and this is sometimes used to clamp down on potential dissidents.

14 See reports by Amnesty International, for instance, *Turkmenistan: Measures of Persuasion*, March 1996; *Kazakhstan: Ill-treatment and the Death Penalty*, July 1996; *Azerbaijan: Time to Abolish the Death Penalty*, March 1997. See also Human Rights Watch, *World Report 1999*, Human Rights Developments.

15 Azerbaijan's flag has, top to bottom, three parallel horizontal bands of blue, red and green, with a white star and crescent moon, horns pointing to the right, in the center of the central band, reminiscent of the Turkish flag. The flag of Kazakhstan has a bright blue background, representing the sky. In the center is a golden sun, wreathed with the outstretched wings of an eagle, resembling ears of corn, and down the left-hand side, a broad vertical band of decorative folk motifs, also in gold. The Turkmen flag is dark green, with a white crescent moon, horns facing to the left, signifying a bright future, and five stars to represent the five

provinces. There is a vertical purple band of traditional carpet designs from the five provinces down the left-hand side, with two crossed olive branches at the bottom.

16 To give just a few examples, in Azerbaijan, the city of Kirovabad, named after a Bolshevik hero, was given back its historical name Gandja; in Baku, Lenin Square was renamed *Azadlig* ("Freedom"). In Turkmenistan, the port of Krasnovodsk was renamed Turkmenbashi; and the provinces of Ashgabat and Charjou were renamed Ahal and Lebap respectively. In Kazakhstan, the port of Shevchenko, named after a Ukrainian poet, was renamed Aktau, while the port of Guriyev became Atyrau.

17 The *manat* in Azerbaijan in August 1992; the *manat* in Turkmenistan and the *tenge* in Kazakhstan in November 1993.

18 For a fuller discussion of Islam during the Soviet period see Shirin Akiner, "Islam, the State and Ethnicity in Central Asia in Historical Perspective," *Religion, State and Society* vol. 24, nos.2/3 (1996): 91–132.

19 The most striking example of Turkmenistan's new rituals is the oath of allegiance, which is repeated everyday in the media, in schools and at all public ceremonies. It runs, in part, "Turkmenistan, beloved fatherland . . . for the slightest calumny about you, may my tongue become powerless; at the moment of treason . . . may my breath be cut off." For the full text, see Michael Ochs, "Turkmenistan: The Quest for Stability," in Dawisha and Parrott, op. cit., 328.

20 Murad Esenov, "Konstitutsiya Türkmenistana," *Tsentral'naya Aziya* no. 2 (1996): 21–25, highlights the ambiguities in the role of the People's Council. In effect, it represents a fourth branch of power, although this is not provided for in the Constitution.

21 See Martha Brill Olcott, "The Growth of Political Participation in Kazakhstan," in Dawisha and Parrott, op. cit., for a fuller account of this period. President Nazarbayev gives his own explanation of events in his autobiography *My Life, My Times and the Future . . .*, translated and edited by Peter Conradi (Northamptonshire: Pilkington Press, 1998), 87–95.

22 Altstadt, op. cit., 129–31.

23 There were many other parties that did not succeed in securing registration. A good account of the positions of different parties is given by Ismailov, op. cit., 36–41.

24 It was registered in spring 1994 and by 1995 claimed to have 64,000 members (Altstadt, op. cit., 148).

25 According to information provided by the First Secretary of the Communist Party, Serikbolsyn Abdildin, in January 1999, it numbered 48,000 members.

26 The deputy chairman of the Movement for Democratic Reform, Khalmurat Soyunov, outlines the problems facing the opposition in Turkmenistan in "Budushcheye Turkmenistana nerazryvna s demokratiyey," *Tsentral'naya Aziya* no. 1, November 1995, 20–25; a similar position is taken by Ashir Ioliyev, "Demokratiya po-Turkmenbashi – svoboda slova uprazdnyayetsya," *Tsentral'naya Aziya* no. 6, 33–37.

27 Turkmen political refugees played a large part in launching the journal *Tsentral'naya Aziya*, later renamed *Tsentral'naya Aziya i Kavkaz*, published in Sweden since 1995.

28 Fuller, op. cit., 144–47.

29 In 1997–98, in Azerbaijan, total active armed forces numbered 66,700, with basic terms of service 17 months; in Kazakhstan, 35,100, with basic terms of service 31 months; in Turkmenistan, approximately 18,000, with terms of service 24 months. (Source: IISS, *The Military Balance 1997/98* [Oxford, Oxford University Press], under relevant country entries.)

30 See Shirin Akiner, "Soviet Military Legacy in Kazakhstan," *Jane's Intelligence Review*, December 1994, 552–54.

31 *Jane's Sentinel: Russia and the CIS, Security Assessment*; 1996 Edition, under entry for Azerbaijan.

32 For an outline of the legal and regulatory framework for the media in these two countries, see Yasha Lange, *Media in the CIS: A Study of the Political, Legislative and Socio-economic Framework* (Dusseldorf: European Institute for Media, 1997), 54–61, 105–109 respectively. Military censorship was introduced in Azerbaijan in 1993, but abolished three years later. Political censorship was introduced in late 1994, but lifted in August 1998.

33 See Lange, op. cit., 213–17.

34 In 1998 it was estimated that 100 newspapers were published in Azerbaijan, including 15 dailies and 30 weeklies. There were two state television channels and three privately owned Baku-based companies. (Source: *Monitoring the Media Coverage of the October 1998 Presidential Elections in Azerbaijan: Final Report*, European Institute for the Media, Dusseldorf, January 1999) As of 1997, there were 234 registered television and radio stations in Kazakhstan, and 725 newspapers (Lange, op. cit., 110). However, most networks reach only a small geographic area. By law, television networks are required to provide 50 percent of their material in Kazakh. In Turkmenistan, as of 1997, there were two television channels one broadcasting in Russian and Turkmen, the other predominantly in Turkmen; there were also 32 periodicals, although only 17 appeared regularly (Lange, op. cit. 218–20).

35 Several such cases are listed in the *Presidential Elections in Azerbaijan: Final Report* (op. cit.), and *Monitoring the Media Coverage of the Kazakhstan 1999 Presidential Elections: Final Report* (Dusseldorf: European Institute for the Media, March 1999).

36 For a good survey of Turkey's hopes, disappointments and achievements in the region see Gareth Winrow, *Turkey in Post-Soviet Central Asia* (London: Royal Institute of International Affairs, 1995).

37 Azeri separatists in Iran, who seek unification with Azerbaijan, are intermittently active. See *RFE/RL Iran Report*, vol. 2, no. 2, January 11, 1999, quoted in *Caghdas Azerbaycan Yolu*, distributed through the Research Centre for Turkestan, Azerbaijan, Crimea, Caucasus and Siberia, SOTA. Haarlem, The Netherlands.

38 Kazakhstan shares a long border with Siberia; much of northeast Kazakhstan is culturally, ethnically and economically more closely linked to Russia than to other parts of Kazakhstan. Azerbaijan borders the Republic of Daghestan, a constituent republic of the Russian Federation. The Lezghi people, one of the main ethnic groups of the Caucasus, are divided between Azerbaijan and Daghestan. From time to time, the Daghestani Lezghis have laid territorial claim to parts of Azerbaijan.

39 This is particularly the case with Kazakhstan and, consequently, good relations with Russia are more of a priority for Kazakhstan than for the other countries. In

May 1992 Kazakhstan signed a bilateral treaty of friendship with Russia; in March 1996, Kazakhstan, along with Russia, Belarus and Kyrgyzstan, signed the "Agreement of Four," forming an inner core of integration within the CIS. The aims of the agreement were to create a united economic area and customs union, to harmonize legal systems, coordinate foreign policy and to provide mutual support in the fight against terrorism.

40 By the beginning of 1995, Aliyev's main political opponents – Ayaz Mutalibov, a former president, Rahim Gaziyev, a former defense minister, and Suret Huseinov, a former prime minister and instigator of a coup against Aliyev in October 1994 – were all based in Moscow. Kazakh opposition leaders also have close links with Moscow, including Akezhan Kazhegeldin, a former prime minister of Kazakhstan and currently the only serious rival to Nazarbayev. There are likewise several Turkmen political dissidents in Moscow. In 1993, a group of them established the Turkmenistan Foundation, under the leadership of the former foreign minister, Abdy Kuliyev. The aims of the Foundation included the annulment of the referendum of January 1994, extending Niyazov's term of office. The Russian authorities generally tolerate such activities; however, they have also on occasion cooperated with the Turkmen government in taking action against the dissidents as, for instance, when they arrested Murad Esenov and Khalmurat Soyunov, leading members of the opposition (Ochs, op. cit., 346–47).

41 Altstadt, op. cit., 130–31; *Presidential Elections in Azerbaijan: Final Report*, op.cit.

42 OSCE/Office for Democratic Institutions and Human Rights (ODIHR), *Election Assessment Mission, Republic of Kazakhstan, January 11, 1999, Preliminary Statement.*

43 In 1997 estimated at 19.3 percent of the labor force in Azerbaijan; 4.1 percent in Kazakhstan (i.e. unemployed registered with the Public Employment Service); recent data not available for Turkmenistan (EBRD, *Transition Report, 1998,* 206–33, under country entries). It is generally accepted, however, that these figures do not reflect the real situation, partly because a very narrow definition of unemployment is used, partly because many out-of-work people do not register with the relevant agencies. The true level of unemployment is widely believed to be very much higher.

44 Recent estimates from officials and research institutes in Kazakhstan put the proportion of those living below the poverty line at 60–75 percent of the population.

45 See Michael Kaser and Santosh Mehrotra, "The Central Asian Economies after Independence," in Allison, op. cit., 287–93.

46 The "mafia" in the FSU have no connection with the mafia in Italy. The term is used loosely to refer to organized criminal groups.

47 For example, the assassination attempt on President Karimov on February 16, 1999, showed that, despite heavy security arrangements, it is difficult to ensure the safety of these leaders.

48 On January 17, 1999, President Aliyev was unexpectedly flown to a military hospital in Ankara for treatment. Official reports stated that he was suffering from bronchitis; unofficial reports hinted at serious heart problems. He returned home on January 30.

49 President Niyazov has suffered from heart problems in the past, but after treatment in Germany in 1997, he has seemed to be in much better health.

50 Saparmurad Niyazov proclaimed a program for a "Decade of Prosperity" in 1992, and in September 1998, Nursultan Nazarbayev put forward his program "2030 Vision." However, in both cases these were general statements of intent rather than concrete plans for reform.

51 Ilham Aliyev is currently Vice-President of SOCAR, the Azeri State Oil Company; there are rumors that he will be made Speaker of the *Mejlis* in the near future, as a prelude, perhaps, to assuming greater responsibility in state affairs. The daughter of Nazarbayev, Dariga, has made a successful career in media management and is currently President of Khabar, the semi-privatized former state television company. The son of the Turkmen president is best known for his profligacy in foreign casinos.

52 In January 1999, Azeri officials were even suggesting that NATO should establish bases in Azerbaijan in order to balance the Russian presence in Armenia (*TransCaucasus: A Chronology*, February 1999).

53 *Kazakhstan: 1998 Country Profile* (London: EBRD). Population estimates are provisional as the full results of the February 1999 census were not available at the time of going to press.

54 *Turkmenistan: 1998 Country Profile* (London: EBRD).

55 *Azerbaijan: 1998 Country Profile* (London: EBRD).

56 For more information on the history of Kazakhstan, see Martha Brill Olcott, *The Kazakhs* (Stanford, CA: Hoover Institution Press, 1987); Shirin Akiner, *The Formation of Kazakh Identity: From Tribe to Nation State* (London: RIIA, 1995).

57 For more information on the history of Azerbaijan, see Tadeusz Swietochowski, *Russian Azerbaijan, 1905–1920: The Shaping of National Identity in a Muslim Community* (Cambridge: Cambridge University Press, 1985); Audrey Altstadt, *The Azerbaijani Turks: Power and Identity Under Russian Rule* (Stanford, CA: Hoover Institution Press, 1992).

5
THE INDUSTRY'S RACE FOR CASPIAN OIL RESERVES

1 Based on operational, not financial data. This includes oil reserves and output, gas reserves and output, refinery capacity and product sales volumes.

2 The one exception is in Azerbaijan, where Section 907 of the 1992 Freedom Support Act restricts official US government aid to Azerbaijan until it stops blockading or using force against neighboring Armenia. The Armenian lobby has succeeded in keeping Section 907 in place, although insurance and lending activities by OPIC and EXIMBANK are now permitted.

3 Eight Western companies are participating in the CPC for a 1.3 million b/d pipeline from the Tengiz field in Kazakhstan to Russia's Black Sea port of Novorissisk – Chevron, Arco through LukArco, Mobil, Shell with Rosneft, BG, Agip, Oryx and BP-Amoco with Kazak Oil. This is in addition to the separate participation of Russia's Lukoil and Rosneft and of the Oman Oil Company (OOC).

6
THE DEVELOPMENT OF CASPIAN SEA OIL: IMPLICATIONS FOR OPEC

1 EIA, *International Energy Outlook* (Washington, DC: US Department of Energy, 1998), 34.
2 IEA, *Caspian Oil and Gas* (Paris: IEA/OECD, 1998).
3 *BP Statistical Review of World Energy*, 1998, 7.
4 Presentation to EIA conference April 1998, as summarized in EIA, *Issues in Midterm Analysis and Forecasting 1998*, 122.
5 IEA, op. cit., 45–61.
6 Ibid., 53.
7 Dr. Matthew Sagers, PlanEcon presentation, Washington, DC, May 1998.
8 EIA, op. cit., 36.
9 Ibid., 25.
10 Ibid., 30–35.
11 US Department of State, speech on "US Policy toward the Caspian Basin" by US Ambassador Richard Morningstar at the Carnegie Endowment, Washington, DC, November 2, 1998.
12 *New York Times*, October 22, 1998 and November 8, 1998.
13 *New York Times*, November 19, 1999.

7
PRICE VS. MARKET SHARE FOR THE ARAB GULF OIL PRODUCERS: DO CASPIAN OIL RESERVES TILT THE BALANCE?

1 Morris Adelman, *Economics of Petroleum Supply* (Cambridge, MA: MIT Press, 1993).
2 Ronald Soligo, Amy Myers Jaffe and Peter Mieszkowski, "Energy Security." Baker Institute for Public Policy, Rice University, 1997, 21–25.
3 Ibid.
4 Dermot Gately, "A Ten Year Retrospective: OPEC and the World Oil Market," *Journal of Economic Literature* vol. 22, no. 2 (1984): 11–14. Also, Dermot Gately, "Strategies for OPEC's Pricing and Output Decisions." *The Energy Journal* vol. 16, no. 3 (1995).
5 Hilliard G. Huntington, "OECD Oil Demand: Estimated Response Surfaces for Nine World Oil Models," vol. 15, no. 1 *Energy Economics*, January 1993, 49–56. Also IEA, "World Energy Outlook," 1994.
6 Gately, "Strategies," op. cit.
7 Edward D. Porter, "Non-OPEC Supply and World Petroleum Markets: Past Forecasts, Recent Experience and Future Prospects," American Petroleum Institute, Washington, DC, 1990.
8 Ibid.

9 See Ronald Soligo and Amy Myers Jaffe, "The Economics of Pipeline Routes: The Conundrum of Oil Exports from the Caspian Basin," Baker Institute for Public Policy, Rice University, 1998, 36–40.
10 *PIW, Oil Market Intelligence* (OMI), December 1998.
11 IEA, "Monthly Oil Market Report," January 1999.
12 Colin J. Campbell and Jean H. Laherrere, "The End of Cheap Oil," *Scientific American*, vol. 278, no. 3, March 1998, 78–85.
13 Michael Lynch, "The Wolf at the Door or Crying Wolf: Fears about the Next Oil Crisis," International Research Center for Energy and Economic Development: 18th International Area Conference, Washington, DC, 1997.
14 Roger N. Anderson, "Recovering Dynamic Gulf of Mexico Reserves and the US Energy Future," *Oil and Gas Journal*, vol. 91, no. 17, April 26, 1993, 85–92.
15 Wood Mackensie Consulting Services, North Sea Reports and South America Service.
16 Knut A. Mork, "Oil and the Macroeconomy When Prices go Up and Down: An Extension of Hamilton's Results," *Journal of Political Economy* vol. 97, no. 3 (1989): 740–744; and Knut A. Mork, Hans T. Mysen and Oystein Olsen, "Macroeconomic Responses to Oil Price Increases and Decreases in Seven OECD Countries," *The Energy Journal*, vol. 15, no. 4 (1994): 19–35.
17 Nathan S. Balke, Stephen Brown and Mine K. Yucel, "Oil Price Shocks and the US Economy: Where Does the Asymmetry Originate?" Presented at AEA meeting, New York, January 1999.
18 Steven J. Davis and John Haltiwanger, "Sectoral Job Creation and Destruction Response to Oil Price Changes and Other Shocks," Paper presented at International Energy Security: Economic Vulnerability to Oil Price Shocks, DOE conference, May 1997.
19 *PIW*, vol. 37, no. 49, April 10, 1989, 1.
20 US Department of Transportation, *Transportation Statistics Annual Report 1997*, 240.
21 Manik Talwani, Andrei Belopolsky and Diane Berry, "Geology and Petroleum Potential of the Caspian Sea Region," Baker Institute for Public Policy, Rice University, 1998.
22 Isabel Gorst and Nina Poussenkova, "Petroleum Ambassadors of Russia: State versus Corporate Policy in the Caspian Region," Baker Institute for Public Policy, Rice University, 1998.
23 Soligo and Jaffe, "Economics," op. cit.

9
THE RELEVANCE OF CASPIAN OIL FOR THE WORLD MARKET

1 This long-term oil outlook examines the global market for crude oil rather than the global market for petroleum products. Measuring crude oil demand against crude oil supply rather than petroleum product demand against crude oil supply provides a better indicator of short-, medium- and long-term crude oil price pressure. Moreover, given the relevance of shifting crude oil trade patterns in assessing the impact of Caspian oil, the focus on crude oil flows is more expedient.

2 *UK Continental Shelf Upstream Oil and Gas Taxation*, Oil and Gas Division Branch 1, UK Department of Trade and Industry, January 1997.
3 For the purposes of this chapter, the Caspian Basin consists of Azerbaijan, Kazakhstan, Turkmenistan and Uzbekistan.
4 The two most well-known independent reviews of Caspian oil resources are *Caspian Oil and Gas: The Supply Potential of Central Asia and Transcaucasia* (Paris: IEA, 1998) and *The World Oil Market*, DOE International Energy Outlook 1998 (Washington, DC: US G.P.O. DOE/EIA-0484(98)).
5 IEA, op. cit., and David I. Hutchison, "Central Asian FSU Republics Set to Expand Oil Export Pipeline Network," *Oil and Gas Journal*, March 9, 1998, 27–33.
6 *The Refining and Distribution System of the Republics of the Former Soviet Union* (Washington, DC: Energy Security Analysis, Inc. 1991), and *Russiawatch* (Washington, DC: ESAI, 1998).
7 Given the uncertainty over pipeline routes, this volume of exports looks increasingly unlikely by 2010. However, it does serve as a plausible long-term objective.
8 This volumetric analysis of the call on OPEC is taken from ESA's proprietary market analysis and forecasting database.
9 *Rules of Thumb for Oil Supply Disruptions*, April 1998, http://www/eia.doe.gov/emeu/security/rule.html.
10 *Tengiz Light: The Way Ahead*, Process Planning Unit, Chevron Research and Technology Company, November 1997.

Bibliography

Adelman, Morris. *Economics of Petroleum Supply* (Cambridge, MA: MIT Press, 1993).

—"Oil Resource Wealth of the Middle East." *Energy Studies Review* vol. 4, no. 1 (1992).

Akiner, Shirin. "Islam, State and Ethnicity in Central Asia in Historical Perspective." *Religion, State and Society* vol. 24, nos. 2/3 (1996).

—*The Formation of Kazakh Identity* (London: RIIA, 1995).

Allison, Roy (ed.) *Challenges for Former Soviet South* (London: RIIA, 1996).

Altstadt, Audrey. *The Azerbaijani Turks: Power and Identity Under Russian Rule* (Stanford, CA: Hoover Institution Press, 1992).

Anderson, Roger N. "Recovering Dynamic Gulf of Mexico Reserves and the U.S. Energy Future." *Oil and Gas Journal*, vol 91, no.17 (April 26, 1993).

"The Art of Doing Business." *Russian Petroleum Investor*, December 1997/January 1998.

"Ashgabat Plans to Award Most of Its Subsoil Licenses for the Year 2000." *Caspian Investor*, September 1998.

"Azerbaijan Attempts to Lure Investors to Its Industrial Sector With Revenue-Sharing Agreements." *Caspian Investor*, June/July 1998.

Azerbaijan Industry Development Initiative. Houston, Texas, August 1998.

"Azerbaijan, Russia Discuss Caspian Sea Status." *Itar-Tass*, March 27, 1998.

Azerbaijan: 1998 Country Profile (London: EBRD, 1998).

Azerbaijan: Time to Abolish the Death Penalty. Amnesty International Report, March 1997.

"Back from the Dead." *Russian Petroleum Investor*, February 1997.

Baker Institute for Public Policy, Rice University. "Unlocking the Assets: Energy and the Future of Central Asia and the Caucasus – A Political, Economic and Cultural Analysis" (April 1998).

"Baku's New Financial Demands May Enable Pipeline Investors to Negotiate Better Terms." *Russian Petroleum Investor*, June/July 1998.

Balke, Nathan S., Stephen Brown and Mine K. Yucel. "Oil Price Shocks and the US Economy: Where Does the Asymmetry Originate?" Presented at AEA meeting, New York, January 1999.

"Better Than Nothing." *Russian Petroleum Investor*, April 1997.

Bohi, Douglas R. "On the Macroeconomic Effects of Energy Price Shocks." *Resources and Energy* vol. 1, no. 2 (June 1991).

—*Analyzing Demand Behavior: A Study of Energy Elasticities* (Baltimore: The Johns Hopkins University Press for Resources of the Future, 1989).

BP Statistical Review of World Energy 1998 (London: The British Petroleum Company, 1998). Also see 1971–1997 issues of this publication.

BP Statistical Review of World Oil Industry 1955–1974 and 1975–1991 (London: The British Petroleum Company).

Campbell, Colin J. "The Next Oil Price Shock: The World's Remaining Oil and Its Depletion." *Energy Exploration and Exploitation* vol. 13, no. 1 (1995).
Campbell, Colin J. and Jean H. Laherrere. "The End of Cheap Oil." *Scientific American*, vol. 278, no. 3, March 1998.
"Caspian Geological Peculiarities Need to be Taken Into Account." *Alexander's Gas & Oil Connections*, April 23, 1998.
"Caspian Investors Contend with Home-Grown Rival." *Russian Petroleum Investor,* June/July 1997.
Caspian Oil and Gas: The Supply Potential of Central Asia and Transcaucasia (Paris: IEA, 1998).
"CCL Wins a Round in Court in its Battle with Astan, but the Fight Continues." *Caspian Investor*, September 1998.
Centre for Global Energy Studies. "Global Oil Report" 1990–1998 issues.
Chorbajian, Levon et al. *The Caucasian Knot: The History and Geo-Politics of Nagorno-Karabagh* (London: Zed Books, 1994).
Chubin, Shahram and Jerrold D. Green. "Engaging Iran: A US Strategy." *Survival* vol. 40, no. 3 (August 1998).
Clagett, Brice. "Rules of International Law." *Caspian Crossroads Magazine* vol. 1, no. 9, December 1996.
Coker, Margaret. "Agreement Could Unlock Caspian." CBS MarketWatch, July 7, 1998.
ConocoEurasia. "The Current Political and Economic Challenges in Azerbaijan, Georgia, Turkmenistan and Kazakhstan," July 1998.
"Cross Caspian Pipeline Would Violate Rights." *Alexander's Gas & Oil Connections,* April 23, 1998.
Davis, Steven J. and John Haltiwanger. "Sectoral Job Creation and Destruction Response to Oil Price Changes and Other Shocks." Paper Presented at Conference on International Energy Security: Economic Vulnerability to Oil Price Shocks, US DOE Conference, May 1997.
Dawisha, Karen and Bruce Parrot (eds) *Conflict, Cleavage and Change in Central Asia and the Caucasus* (Cambridge: Cambridge University Press, 1997).
"Deeper Pockets." *Russian Petroleum Investor*, December 1997/January 1998.
DeLay, Jennifer quoted in *Russia Today*, June 8, 1998.
Denton Hall, Doing Business in Azerbaijan (London: Denton Hall International Group, 1998)
DIW. "Die Energiewirtshaft am Kaspischen Meer: Entauschte Erwartungen-Unischere Perspektiven," *DIW Wochenbericht* vol. 24, no. 98, June 11, 1998.
Ebel, Robert E. *Energy Choices in the Near Abroad: The Haves and the Have-nots Face the Future* (Washington, DC: CSIS, 1997).
"ECO Foreign Ministers Met." *Alexander's Gas & Oil Connection*, May 11, 1998.
"Economic, Not Political, Reforms Are More Likely to Ensure Contract Stability in Turkmenistan." *Russian Petroleum Investor*, June/July 1998.
Embassy of Turkmenistan. Statement by Ministry of Foreign Affairs, July 7, 1998.
Energy Information Agency. *International Energy Outlook 1998* (Washington, DC: US Department of Energy, April 1998).
—*Issues in Midterm Analysis and Forecasting 1998* (Washington, DC: US DOE, 1998).
Energy Security Analysis, Inc. *The Refining and Distribution System of the Republics of the Former Soviet Union* (Washington, DC: Energy Security Analysis, Inc., 1991).

Entezari, Shirin and Aleen Rothschild-Seidel. "Azerbaijan: Laws Concerning Privatization and Foreign Investment." *Caspian Crossroads Magazine* vol. 1, no. 3 (Summer/Fall 1995).
Erzysbayev, Ermukhamet. "Nekotoryye aspekty prezidentskoy izbiratel'noy kampanii v Kazakhstane." *Tsentral'naya Aziya i Kavkaz* vol. 1, no. 3 (1999).
Esenov, Murad. "Konstitutsiya Turkmenistana." *Tsentral'naya Aziya* no. 2 (1996).
European Commission Directorate-General XVII (Energy). The Energy Charter Treaty, September 15, 1997.
Fischer, Louis. *Oil Imperialism* (New York: International Publishers, 1926).
"For Three Joint Ventures, Tax Liability Expands Beyond Expectations." *Russian Petroleum Investor*, June/July 1998.
"Foreign Investors May Finally get Access to Turkmenistan's Offshore Deposits." *Russian Petroleum Investor*, June/July 1997.
Forsythe, Rosemarie. *The Politics of Oil in the Caucasus and Central Asia: Prospects for Oil Exploitation and Export in the Caspian Basin.* Adelphi paper no. 350 (London: IISS, 1996).
Fuller, Graham. "Geopolitical Dynamics of the Caspian Region." *Caspian Crossroads Magazine* vol. 3, no. 2 (Fall 1997).
Fyodorov, Yuri, Rajan Menon and Ghia Nodia (eds) *Russia, the Caucasus and Central Asia: The 21st Century Security Environment* (Armonk, NY: M.E. Sharpe, 1999).
Gately, Dermot. "A Ten Year Retrospective: OPEC and the World Oil Market." *Journal of Economic Literature* vol. 22, no. 2 (1984).
—"Lessons from the 1986 Oil Price Collapse." *Brookings Papers on Economic Activity* vol. 2, (1986).
—"Strategies for OPEC's Pricing and Output Decisions." *The Energy Journal* vol. 16, no. 3 (1995).
Gorst, Isabel and Nina Poussenkova. "Petroleum Ambassadors of Russia: State versus Corporate Policy in the Caspian Region." Baker Institute for Public Policy, Rice University, April 1998.
"Government to Propose Liberalizing Rules for Transfer of Ownership and Sub-Soil Use Licenses." *Russian Petroleum Investor*, June/July 1998.
Gregory, Paul. *Reforming the Soviet Economic Bureaucracy* (Cambridge: Cambridge University Press, 1983).
—"The Russian Energy Sector: Corporate Governance, Privatization and Financing Capital Accumulation." Paper Presented at a Conference on the Energy Sector in the Former Soviet Union, Oxford University, St. Anthony's College, March 20–21, 1998.
Griffen, James B. and David J. Teece (eds) *OPEC Behavior and World Oil Prices.* (London: George Allen and Unwin Ltd., 1982) 145–73.
Gustafson, Thane. *Crisis Among Plenty: The Politics of Soviet Energy Under Brezhnev and Gorbachev* (Princeton, NJ: Princeton University Press, 1989).
Gustafson, Thane and Laurent Ruseckas. "The Future of Caspian Oil: Can a 'Great Game' be Averted?" CERA (December 1997).
Heslin, Sheila. "Key Constraints to Caspian Pipeline Development: Status, Significance, and Outlook." Baker Institute for Public Policy (April 1998).
Hirschhausen, C. and H. Engerer. "Gas Sector Restructuring in the CIS: A Political Economy Approach." DIW Berlin, February 1998.

Hosseini, A. H. "The Prospects for the Caspian Sea Region in the 21st Century." Institute for Political and International Studies, Summer/Fall, 1997.
Human Rights Watch/Helsinki. *Azerbaijan: Seven Years of Conflict in Nagorno-Karabakh* (New York, NY: Human Rights Watch, 1994).
Huntington, Hillard G. "OECD Oil Demand: Estimated Response Surfaces for Nine World Oil Models." *Energy Economics*, vol. 15, no. 1 (January 1993).
Ioliyev, Ashir. "Demokratiya po-Turkmenbashi – Svoboda Slova Uprzdnyayetsya." *Tsentral'naya Aziya* no. 6 (1996).
International Monetary Fund; International Financial Statistics Yearbook (Washington DC: IMF, 1997).
International Monetary Fund Monthly Review, June 1997.
International Energy Agency. *Caspian Oil and Gas* (Paris: IEA/OECD, 1998).
—*World Energy Outlook, 1998 Edition* (Paris: International Energy Agency, 1998).
—"Monthly Oil Report and World Energy Outlook," December 1998, January 1999.
"Iran Attacks Caspian Oil Deal." BBC News Online Asia-Pacific, July 8, 1998.
"Iran Increasingly Isolated on Caspian Argument." *Alexander's Gas & Oil Connections*, May 4, 1998.
"Irate Iranians." *Russian Petroleum Investor*, December 1997/January 1998.
Ismailov, El'bar. "Vlast' I Oppozitsiya Nakanune i v Period Prezidentskoy Izbiratel'noy Kampanii v Azerbaydzhane." *Tsentral'naya Aziya i Kavkaz* vol. 1, no. 2 (1999).
Jaffe, Amy Myers and Robert Manning. "The Myth of the Caspian 'Great Game:' The Real Geopolitics of Energy," *Survival* vol. 40, no. 4 (Winter 1998–99).
Kalyuzhnova, Yelena. *The Kazakhstani Economy: Independence and Transition* (London: Macmillan, 1998).
"KazakOil's Ambitious Growth Goals Offer New Opportunities to Investors." *Russian Petroleum Investor*, June/July 1998.
"Kazakh Oil and Gas Reserves Are Waiting." *Alexander's Oil & Gas Connection*, April 30, 1998.
"Kazakhstan Betting on Several Horses." *Alexander's Gas & Oil Connections*, May 12, 1998.
Kazakhstan: 1998 Country Profile (London: EBRD, 1998).
Kazakhstan: Ill-treatment and the Death Penalty. Amnesty International Report, July, 1996.
Kemp, Geoffrey. *America and Iran: Road Maps and Realism* (Washington, DC: The Nixon Center for Peace and Freedom, 1998).
—*Energy Superbowl: Strategic Politics in the Persian Gulf and the Caspian Basin* (Washington, DC: Nixon Center for Peace and Freedom, 1997).
Kemp, Geoffrey and Robert Harkavy. *Strategic Geography and the Changing Middle East* (Washington, DC: The Brookings Institution Press, 1997).
Lange, Yashe. *Media in the CIS: A Study of the Political, Legislative and Socio-economic Framework* (Dusseldorf : European Institute for Media, 1997).
"Licensed to Drill." *Russian Petroleum Investor*, June/July 1997.
Lynch, Michael C. "Facing the Elephant: Oil Market Evolution and Future Oil Crises." International Research Center for Energy and Economic Development, Occasional Paper, no. 30, 1998.
—"The Wolf at the Door or Crying Wolf: Fears about the Next Oil Crisis." International Research Center for Energy and Economic Development: 18th International Area Conference, Washington, DC, 1997.
Mabro, Robert. "OPEC and the Price of Oil." *The Energy Journal* vol. 13, no. 2 (1990).

Menon, Rajan. "Treacherous Terrain: The Political and Security Dimensions of Energy Development in the Caspian Sea Zone." *NBR Analysis* vol. 9, no. 1 (February 1998).
Momtaz, Jamshid. "The Legal Regime of the Caspian Sea." *Amu Darya* vol. 1, no. 2 (Summer/Fall 1996).
Mork, Knut A. "Oil and the Macroeconomy when prices go up and down: an extension of Hamilton's Results." *Journal of Political Economy* vol. 97, no. 3 (1989).
Mork, Knut A., Hans T Mysen and Oystein Olsen. "Macroeconomic Responses to Oil Price Increases and Decreases in Seven OECD Countries." *The Energy Journal* vol. 15, no. 4 (1994).
Morningstar, Richard L. "Testimony before the Senate Subcommittee on International Economic Policy, Exports, and Trade Promotion," March 3, 1999.
"Much Ado About Nothing." *Russian Petroleum Investor*, April 1998.
Mueller, F. "Die Region des Kaspischen Meeres Energiereichtum und Geopolitik." *Osteuropa Wirtshaft* vol. 41, no. 3.
Nazarbayev, Nursultan. *My Life, My Times and the Future* (translated and edited by Peter Conradi) (Northamptonshire: Pilkington Press, 1998).
"New and Improved." *Russian Petroleum Investor*, February 1997.
Nordhaus, William D. "Oil and Economic Performance in Industrial Countries." *Brookings Papers on Economic Activity* vol. 2 (1980).
Occidental of Umm Al Qaywayn, Inc. v. *A Certain Cargo of Petroleum*, 577 F.2d, 1196 (5th Circuit, 1978).
Oil and Gas Journal (Tulsa: Pennwell Publishing Co., selected articles: 1996–1998).
Olcott, Martha Brill. "The Caspian's False Promise." *Foreign Policy* no. 111 (Summer 1998).
Order of the President of Turkmenistan. no. 3164, May 27, 1997.
OPEC. *Annual Statistical Bulletin*, 1951, 1955, 1957, 1970–1997 (Vienna).
OSCE, ODIHR. Election Assessment Mission, Republic of Kazakhstan. Preliminary Statement, January 11, 1999.
Oxman, Bernard. "Caspian Sea or Lake: What Difference Does It Make?" *Caspian Crossroads Magazine* vol. 1, no. 4 (Winter 1996).
"The Perfect Host." *Russian Petroleum Investor*, April 1998.
Petroleum Industry Research Associates (PIRA). Report to Retainer Clients (1996, 1997, 1998).
Petroleum Intelligence Weekly, various articles, *Oil Market Intelligence*, Energy Intelligence Group, 1988–1998.
"Pipeline Dreams: The Struggle for Caspian Oil." *Washington Post*, October 6, 1998.
PlanEcon. *Energy Outlook for the Former Soviet Republics*, June 1995.
"Planned Pipelines Would Open Turkmenistan to New Lucrative Markets." *Russian Petroleum Investor*, June/July 1997.
Platt's Oilgram News (New York: Mcgraw Hill Companies, selected articles: 1997–1998).
Pomfret, Richard. "The Economic Cooperation Organization: Current Status and Future Prospects." *Europe–Asia Studies* vol. 49, no. 4 (June 1997).
Porter, Edward D. "Non-OPEC Supply and World Petroleum Markets: Past Forecasts, Recent Experience and Future Prospects." API, Washington, DC, 1990.
"Pro-PSA Deputies Plan to Enact PSA Laws This Summer." *Russian Petroleum Investor*, June/July 1998.
"Progress for the CPC Does Not Necessarily Mean Progress for Oil Producers in Kazakhstan." *Russian Petroleum Investor*, June/July 1998.

Roberts, John. *Caspian Pipelines* (London: RIIA, 1996).
Ruseckas, Laurent. "State of the Field Report: Energy and Politics in Central Asia and the Caucasus." *AccessAsia Review* vol. 1, no. 2 (July 1998).
"Russia. Kazakhstan Sign Oil Pact, Ending Protracted Row." BT, July 7, 1998.
"Russia/Kazakhstan: Yeltsin, Nazarbayev Ink Accord on Caspian Oil Rights." *RFE/RL*, July 6, 1998.
Russiawatch (Washington, DC: ESAI, selected issues: 1997–1998).
Shenoy, Bhamy and S. Gurcan Gulen. *Caspian Pipeline Game.* Energy Institute, College of Business Administration, University of Houston, 1998.
—*Politics and Economics of Proposed Caspian Export Pipeline Projects*, Current Studies on Business/Public Policy Issues in Energy, An Energy Institute White Paper, December 1998.
Shimizu, Manabu (ed.) *The Caspian Basin Oil and its Impact on Eurasian Power Games* (Tokyo: IDE, IDE Spot Survey, June 1998).
Shoumikhin, Andrei. "New Developments Related to Caspian Oil," Proceedings of the International Workshop *Caspian Oil: Transcaucasus, Central Asia and International Security*, Moscow Public Science Foundation, vol. 1, no.9, December 1996.
Skagen, Ottar. *Caspian Gas* (London: RIIA, 1997).
Slone, Daniel and John Lain. "Open For Business: The Legislative Framework of the Republic of Kazakhstan." *Caspian Crossroads Magazine* vol. 1, no. 3 (Summer/Fall 1995).
Soligo, Ronald, Amy Myers Jaffe, Peter Mieszkowski. *Energy Security*. Baker Institute for Public Policy, Rice University, 1997.
Soligo, Ronald and Amy Myers Jaffe. *The Economics of Pipeline Routes: The Conundrum of Oil Exports from the Caspian Basin.* (Texas: Baker Institute for Public Policy, Rice University, 1998).
—*Unlocking the Assets: Energy and the Future of Central Asia and the Caucasus – A Political, Economic and Cultural Analysis* (Texas: Baker Institute for Public Policy, Rice University, April 1998).
Soyunov, Khalmurat. "Budushcheye Turkmenistana nerazryvna s demokratiyey." *Tsentral'naya Aziya* no. 1 (1995).
Starr, Frederick. "Power Failure: American Policy in the Caspian." *National Interest* no. 47 (Spring 1997).
Swietochowski, Tadeusz. *Russian Azerbaijan, 1905–1920: The Shaping of National Identity Under Russian Rule* (Cambridge: Cambridge University Press, 1985).
Talwani, Manik, Andrei Belopolsky and Diane Berry. *Geology and Petroleum Potential of the Caspian Sea Region.* (Baker Institute for Public Policy, Rice University, 1998.
"Ten Caspian States Meet and Talk." *Alexander's Gas & Oil Connections*, May 11, 1998).
Tolf, Robert W. *The Russian Rockefellers* (Standford, CA: Standford University, Hoover Institution Press, 1976).
Transition Report 1998 (London: EBRD, 1998).
Turkmenistan: 1998 Country Profile (London: EBRD, 1998).
Turkmenistan: Measures of Persuasion. Amnesty International Report, March 1996.
"Turkmenistan and Azerbaijan Clash Over Caspian Sea." *Alexander's Gas & Oil Connection*, January 30, 1997.
"Unclogging the Pipe." *Russian Petroleum Investor*, December 1997/January 1998.

US Department of Energy. *The World Oil Market*, DOE International Energy Outlook 1998 (Washington, DC: US GPO DOE/EIA-0484(98)).

US Department of State. "Caspian Region Energy Development Report as Required by HR 3610." Congressional Record, May 8, 1997.

—"US policy toward the Caspian Basin." Speech by US Ambassador Richard Morningstar at the Carnegie Endowment, Washington, DC, November 2, 1998.

US Department of Transportation, Washington, DC, *Transportation Statistics Annual Report 1997*.

Waelde, Thomas W. and Christian von Hirschhausen. "Regulatory Reform in the Energy Industry of Post-Soviet Countries," *DIW Discussion Paper* no. 155, November 1997

Winrow, Gareth. *Turkey in Post-Soviet Central Asia* (London: Royal Institute of International Affairs, 1995) OSCE/ODIHR. Election Assessment Mission, Republic of Kazakhstan. Preliminary Statement, January 11, 1999.

World Bank. Strategic Export Expansion Study, *Interim Report*, July 22, 1996.

Yergin, Daniel and Thane Gustafson. "Evolution of an Oil Rush," *New York Times*, August 6, 1997.

Index